Studies in Computational Intelligence

Volume 569

Series editor

Janusz Kacprzyk, Polish Academy of Sciences, Warsaw, Poland
e-mail: kacprzyk@ibspan.waw.pl

About this Series

The series "Studies in Computational Intelligence" (SCI) publishes new developments and advances in the various areas of computational intelligence—quickly and with a high quality. The intent is to cover the theory, applications, and design methods of computational intelligence, as embedded in the fields of engineering, computer science, physics and life sciences, as well as the methodologies behind them. The series contains monographs, lecture notes and edited volumes in computational intelligence spanning the areas of neural networks, connectionist systems, genetic algorithms, evolutionary computation, artificial intelligence, cellular automata, self-organizing systems, soft computing, fuzzy systems, and hybrid intelligent systems. Of particular value to both the contributors and the readership are the short publication timeframe and the world-wide distribution, which enable both wide and rapid dissemination of research output.

More information about this series at http://www.springer.com/series/7092

Roger Lee

Editor

Software Engineering, Artificial Intelligence, Networking and Parallel/Distributed Computing

 Springer

Editor
Roger Lee
Software Engineering & Information
    Technology Institute
Central Michigan University
USA

ISSN 1860-949X   ISSN 1860-9503   (electronic)
ISBN 978-3-319-36173-4   ISBN 978-3-319-10389-1   (eBook)
DOI 10.1007/978-3-319-10389-1

Springer Cham Heidelberg New York Dordrecht London

Printed on acid-free paper

Springer is part of Springer Science+Business Media (www.springer.com)

Preface

The purpose of the 15th IEEE/ACIS International Conference on Software Engineering, Artificial Intelligence, Networking and Parallel/Distributed Computing (SNPD 2014) held on June 30 – July 2, 2014 in Las Vegas Nevada, USA, is aimed at bringing together researchers and scientists, businessmen and entrepreneurs, teachers and students to discuss the numerous fields of computer science, and to share ideas and information in a meaningful way. This publication captures 13 of the conference's most promising papers, and we impatiently await the important contributions that we know these authors will bring to the field.

In chapter 1, Junping Song, Haibo Wang, Pin Lv, Shangzhou Li, and Menglu Xu propose a data mining based publish/subscribe system (DMPSS). The experimental results show that DMPSS realizes even matching load distribution, and it reduces the overhead for message transmission and latency dramatically.

In chapter 2, Umme Habiba, Rahat Masood and Muhammad Awais Shibli propose a secure identity management system for federated Cloud environments that not only ensures the secure management of identity credentials, but preserves the privacy of Cloud Service Consumers (CSC) also. The results of their evaluation certify that the presented work ensures the desired features and level of security as expected from a secure identity management system for federated Cloud environment.

In chapter 3, Abdulmohsen Al-Thubaity, Muneera Alhoshan, and Itisam Hazzaa conducted studies on the effect of using word N-grams (N consecutive words) on ATC accuracy. Their results show that the use of single words as a feature provides greater classification accuracy (CA) for ATC compared to N-grams.

In chapter 4, Mohammad Hadi Valipour, Khashayar Niki Maleki and Saeed Shiry Ghidary propose a new approach in order to control unstable systems or systems with unstable equilibrium. Evaluation measures in simulation results show the improvement of error reduction and more robustness than a basic tuned double-PID controller for this task.

In chapter 5, Yadu Gautam, Carl Lee , Chin-I Cheng, and Carl Langefeld study the performance of the MiDCoP approach using association analysis based on the imputed allele frequen-cy by analyzing the GAIN Schizophrenia data. The results indicate that the choice of reference sets has strong impact on the performance.

In chapter 6, Soumya Saha and Lifford McLauchlan propose an approach to construct weighted minimum spanning tree in wireless sensor networks. Simulation results demonstrate significant improvement for both load balancing and number of message deliveries after implementation of the proposed algorithm.

In chapter 7, Taku Jiromaru, Tetsuo Kosaka, and Tokuro Matsuo collected answers of each learn-er for knowing element of difficulty level in Mathematics, and identified 10 types.

In chapter 8, Lifeng Zhang, Akira Yamawaki and Seiichi Serikawa propose an approach to identify and exter-minate a specialized invasive alien fish species, the black bass. Simulation result shows a reasonable possibility for identify a black bass from other fish species.

In chapter 9, Nermin Kajtazovic, Christopher Preschern, Andrea Holler and Christian Kreiner present a novel approach for verification of compositions for safety-critical systems, based on data semantics of components. They show that CSP can be successfully applied for verification of compositions for many types of properties.

In chapter 10, Mitsuo Wakatsuki, Etsuji Tomita, and Tetsuro Nishino study a subclass of deterministic pushdown transducers, called deterministic restricted one-counter transducers (droct's), and studies the equivalence problem for droct's which accept by final state.

In chapter 11, Hiroki Nomiya, Atsushi Morikuni, and Teruhisa Hochin propose a emotional scene detection in order to retrieve impressive scenes from lifelog videos. The detection performance of the proposed method is evaluated through an emotional scene detection experiment.

In chapter 12, Takafumi Nakanishi presents a new knowledge extraction method on Big Data Era. In this paper, he especially focus on an aspect of heterogeneity. He discovers a correlation in consideration of the continuity of time.

In chapter 13, Golnoush Abaei and Ali Selamat propose a new method is proposed to increase the accuracy of fault prediction based on fuzzy clustering and majority ranking. The results show that their systems can be used to guide testing effort by prioritizing the module's faults in order to improve the quality of software development and software testing in a limited time and budget.

It is our sincere hope that this volume provides stimulation and inspiration, and that it will be used as a foundation for works to come.

Guest Editors

June 2014   Ju Yeon Jo
University of Nevada-Las Vegas, U.S.A
Satoshi Takahashi
University of Electro-Communications, Japan

Contents

List of Contributors

Golnoush Abaei
University Technology Malaysia,
   Malaysia
golnoosh.abaee@gmail.com

Abdulmohsen Al-Thubaity
KACST, Saudi Arabia
aalthubaity@kacst.edu.sa

Muneera Alhoshan
KACST, Saudi Arabia
malhawshan@kacst.edu.sa

Chin-I Cheng
Central Michigan University, USA
cheng3c@cmich.edu

Yadu Gautam
Central Michigan University, USA
gauta1yn@cmich.edu

Umme Habiba
School of Electrical Engineering and
   Computer Science, Pakistan
11msccsuhabiba@seecs.edu.pk

Itisam Hazzaa
KACST, Saudi Arabia
429202034@student.ksu.
edu.sa

Teruhisa Hochin
Kyoto Institute of Technology, Japan
hochin@kit.ac.jp

Andrea Holler
Graz University of Technology, Austria
andrea.hoeller@tugraz.at

Taku Jiromaru
Yamagata University, Japan
jiro@om-edu.jp

Nermin Kajtazovic
Graz University of Technology, Austria
nermin.kajtazovic@tugraz.at

Tetsuo Kosaka
Yamagata University, Japan
tkosaka@yz.yamagata-u.ac.jp

Christian Kreiner
Graz University of Technology, Austria
christian.kreiner@tugraz.at

Carl Langefeld
Wake Forest University, USA
clangefe@wfubmc.edu

Carl Lee
Central Michigan University, USA
carl.lee@cmich.edu

Shangzhou Li
ISCAS, China
lishangzhou2012@sina.com

Pin Lv
ISCAS, China
lvpin@iscas.ac.cn

Rahat Masood
School of Electrical Engineering and
    Computer Science, Pakistan
rahat.masood@seecs.edu.pk

Tokuro Matsuo
Advanced Institute of Industrial
    Technology, Japan
matsuo@tokuro.net

Lifford McLauchlan
Texas A&M University, USA
lifford.mclauchlan@tamuk.
edu

Atsushi Morikuni
Kyoto Institute of Technology, Japan
m2622043@edu.kit.ac.jp

Takafumi Nakanishi
International University of Japan, Japan
takafumi@glocom.ac.jp

Khashayar Niki Maleki
University of Tulsa, Iran
kh.niki.m@gmail.com

Hiroki Nomiya
Kyoto Institute of Technology, Japan
nomiya@kit.ac.jp

Christopher Preschern
Graz University of Technology, Austria
christpher.preschern@
tugraz.at

Soumya Saha
Texas A&M University, USA
jishumail@gmail.com

Ali Selamat
University Technology Malaysia,
    Malaysia
aselamat@utm.my

Seiichi Serikawa
Kyushu Institute of Technology, Japan
serikawa@elcs.kyutech.ac.jp

Muhammad Awais Shibli
School of Electrical Engineering and
    Computer Science, Pakistan
awais.shibli@seecs.edu.pk

Saeed Shiry Ghidary
Amirkabir University of Technology,
    Iran
shiiry@aut.ac.ir

Junping Song
ISCAS, China
junping@iscas.ac.cn

Etsuji Tomita
The University of Electro-
    Communications, Japan
tomita@uec.ac.jp

Mohammad Hadi Valipour
Amirkabir University of Technology
    Iran
valipour@aut.ac.ir

Mitsuo Wakatsuki
The University of Electro-
    Communications, Japan
wakatsuki.mitsuo@uec.ac.jp

Haibo Wang
ISCAS, China
haibo@iscas.ac.cn

Menglu Xu
ISCAS, China
lumengxu@gmail.com

Akira Yamawaki
Kyushu Institute of Technology, Japan
yama@elcs.kyutech.ac.jp

Lifeng Zhang
Kyushu Institute of Technology, Japan
zhang@elcs.kyutech.ac.jp

A Data Mining Based Publish/Subscribe System over Structured Peer-to-Peer Networks

Junping Song, Haibo Wang, Pin Lv, Shangzhou Li, and Menglu Xu

Abstract. In this paper, we propose a data mining based publish/subscribe system (DMPSS). First, the data mining technology is used to find attributes that are usually subscribed together, e.g. frequent itemset. Then subscriptions and events are installed by frequent itemsets contained in them. If subscriptions and events don't contain any frequent itemset, they are delivered to specified RPs (rendezvous points) for matching. The usage of frequent itemsets provides two advantages to DMPSS. First, it achieves even matching load distribution on RPs. Second, it reduces the event publication cost. The performance of DMPSS is evaluated by simulations. The experimental results show that DMPSS realizes even matching load distribution, and it reduces the overhead for message transmission and latency dramatically.

1 Introduction

The publish/subscribe system is widely used for delivering data from publishers (data producers) to subscribers (data consumers) across large-scale distributed networks in a decoupled fashion. Traditional publish/subscribe applications include news and sports ticker services, real-time stock quotes and updates, market tracker, etc. [1] In recent years, the publish/subscribe scheme has been used in many Web 2.0 applications, especially On-line Social Networks like Twitter, Facebook and Google+ [2]. However, most of the above mentioned applications adopt topic-based subscriptions, which offer very limited expressiveness to subscribers. Currently, the study of content-based systems, which allow fine-grained subscriptions by enabling restrictions on the event content, has attracted plenty of attention [3-7].

Junping Song · Haibo Wang · Pin Lv · Shangzhou Li · Menglu Xu
Science and Technology on Integrated Information System Laboratory,
Institute of Software, Chinese Academy of Sciences, 4# South Fourth Street,
Zhong Guan Cun, Beijing 100190 P.R. China
e-mail: {junping,haibo,lvpin}@iscas.ac.cn,
   lishangzhou2012@sina.com, lumengxu@gmail.com

© Springer International Publishing Switzerland 2015   1
R. Lee (ed.), *SNPD*,
Studies in Computational Intelligence 569, DOI: 10.1007/978-3-319-10389-1_1

With the development of distributed technologies, Distributed Hash Table (DHT) is widely employed to provide systems with scalability and self-organizing. In a DHT-based publish/subscribe system, a node can play three roles simultaneously, including subscriber, publisher and broker. As a subscriber, it registers its interest on some brokers by means of subscriptions, and receives events that satisfy its interest from brokers. As a publisher, it sends events that contain certain information to brokers, regardless of the final receivers. And as a broker, it matches received events with subscriptions and delivers them to the matched subscribers.

In a DHT-based publish/subscribe system, a broker is also called a rendezvous point (RP). Generally speaking, only when an event and a subscription meet on a RP can they match. As a result, the policy for subscription installation and event publication is of critical importance. Up to now, several proposals have been put forward for DHT and content based systems. Gupta [3] and Li [8] create Cartesian spaces whose dimensions are decided by the number of attributes in the system. Each RP is responsible for the matching within a sub-space, and subscriptions and events are installed according to value ranges or values of their attributes. Complexities of such systems increase dramatically with the growth of the number of attributes in the system. The proposals of Terpstra [4] and Silvia [5] apply filter-based routing, and they combine the matching process with event publication. However, organization and maintenance of subscriptions are always complicated for filter-based routing strategy. Ferry [6] and Eferry [7] manage subscriptions and events according to names of attributes contained in them. Ferry assigns a subscription to a certain RP by hashing the name of an attribute contained in the subscription, and each event is published to all RPs for matching. Eferry improves Ferry to maintain a suitable quantity of RP nodes in the system and keep an even load distribution among them, but it suffers from high event publication cost. PEM [9] and Vinifera [10] also install their subscriptions according to names of attributes.

The loads of RPs and the corresponding load balancing performance are always key issues in DHT-based publish/subscribe systems. In fact, the load of a RP mainly comprises two parts: overhead for matching and overhead for message transmission. The former is affected by three factors: the number of subscriptions stored on the RP, the number of events the RP receives and the matching complexity. And the latter is impacted by the number of subscriptions and events exist in the system. However, none of the previous proposals evaluated the loads of RPs comprehensively.

In this paper, we propose a data mining based publish/subscribe system (DMPSS). In DMPSS, multiple factors are taken into consideration to optimize loads of RPs. We use data mining technology to find attributes that are usually subscribed together, e.g. frequent itemset. Then RPs are divided into two categories. One category is responsible for matching subscriptions and events that contain frequent itemsets, and the other category is responsible for matching subscriptions and events that don't contain frequent itemsets. Subscriptions and events are installed by frequent itemsets contained in them. The advantages of DMPSS are as follows:

1. It can be applied on any overlay networks easily.

2. It achieves even matching load distribution on RPs. In DMPSS, RPs related to frequent itemsets store larger number of subscriptions than RPs that are not related

to frequent itemsets. However, the numbers of events that are sent to RPs related to frequent itemsets are less compared with the numbers of events that are sent to the other category of RPs. As a result, the total matching times on RPs are even.

3. It reduces the event publication cost. DMPSS groups subscriptions that do not contain frequent itemsets onto a certain number of RPs. As a result, the numbers of events that are sent to RPs for matching are reduced, and the overheads of nodes for message transmission are reduced correspondingly.

The remainder of this paper is structured as follows. Section 2 introduces the background and key technologies that are used in DMPSS. Section 3 presents the proposed DMPSS scheme. In Section 4, we evaluate the performance of DMPSS by simulations. And Section 5 offers some concluding remarks.

2 Background

2.1 System Definition

Similar to most of the previous publish/subscribe systems, DMPSS is defined as $S = \{A_1, A_2, \ldots, A_n\}$, where A_i represents an attribute. Each attribute is described by a tuple $[name : type, min, max]$. A subscription is a conjunction of predicates over the attribute values, i.e., $sub = p_1 \wedge \cdots \wedge p_m$, where p_i specifies a constant value or a range for an attribute using common operators ($=, >, \geq, <, \leq, \neq$, etc.). For example, a subscription expressed on the attribute A_1 and A_2 may be of the form $sub = (a < A_1 < b) \wedge (c < A_2 < d)$. An event is a set of equalities on attributes, e.g., $e = \{\ldots, A_i = v_i, \ldots\}$. In this paper, if the value of an attribute is not set in an event, we suppose it is NULL.

In this paper, the attribute set that appears in a subscription sub is defined as $Attr_{sub}$, and that appears in an event e is defined as $Attr_e$. For example, when $sub_1 = (a < A_1 < b) \wedge (c < A_2 < d)$ and $e_1 = \{A_1 = v_1, A_2 = v_2, A_3 = v_3\}$, $Attr_{sub1} = \{A_1, A_2\}$ and $Attr_{e1} = \{A_1, A_2, A_3\}$. For ease of description, a subscription sub or an event e contains attribute set $attr$ means $attr \subseteq Attr_{sub}$ or $attr \subseteq Attr_e$. An event e matches a subscription sub if and only if each predicate of sub is satisfied by the value of corresponding attribute contained in event e. That is to say, if an event e matches a subscription sub, $Attr_{sub} \subseteq Attr_e$.

2.2 DHT-Based Publish/Subscribe System

The DHT overlay network is a class of decentralized systems in the application level that partition ownership of a set of objects among participating nodes, and can efficiently route message to the unique owner of any given object [1]. The classic DHT overlay networks include Chord [11], CAN [12], Tapestry [13], Pastry [14], etc. Each object or node in a DHT overlay network is assigned an ID (i.e. key) according to a consistent hash function. An object is stored in a node whose ID closest or immediately succeeds to the object's ID.

UP to now, several DHT-based publish/subscribe systems have been proposed, including topic-based [15-16] and content-based [2-4] systems. Most of them are designed on the basis of existing DHT networks, and the other proposals are put forward based on novel DHT structures [5].

In this paper, we propose a DHT-based publish/subscribe system that can be applied on any DHT overlay networks. In DMPSS, we only provide methods that specify where subscriptions are stored and where events are published to. Users can choose the overlay network according to their needs. And the routing mechanism of the message is decided by the overlay that the user chooses.

2.3   Frequent Itemset Mining

Frequent itemsets are itemsets that appear in a data set with frequency no less than a user-specified threshold. For example, a set of items, such as milk and bread, that appears frequently together in a transaction data set, is a frequent itemset. Given a itemset database D, the support of a itemset α is the percentage of itemsets in D which contain α. Let $I = \{i_1, i_2, \ldots, i_n\}$ be a set of all items. A k-itemset α, which consists of k items from I, is frequent if the support of α no lower than a user-specified minimum support threshold [17].

In DMPSS, we extract attribute sets from subscriptions and find frequent itemsets using frequent itemset mining method, such as FP-growth [18] and Apriori [19]. Then we install subscriptions and events in terms of frequent itemsets. The details of DMPSS are illustrated in the next section.

3   System Design

The usage of frequent itemset for subscription and event installation is the most important innovation of DMPSS. The frequent itemset based policy of DMPSS realizes two advantages. First, it achieves even matching load distribution on RPs. Second, it reduces the event publication cost by grouping subscriptions that don't contain frequent itemsets onto a certain number of RPs.

In this section, details of DMPSS are presented in Section 3.1 - Section 3.6. Section 3.7 introduces a load balancing policy.

3.1   Frequent Itemsets Mining from Historical Recorders of Subscriptions

At first, we extract attribute sets from historical records of subscriptions. Then set the minimum support threshold $minsup$ and find the frequent itemsets using frequent itemset mining method.

In order to management subscriptions that don't contain frequent itemset, we select M non-frequent itemset randomly, and call them virtual frequent itemset. The value of M is limited by

$$M \geq \frac{1 - S_{NF}}{\max load}$$

where S_{NF} is the percentage of attribute sets which contain frequent itemset, and *maxload* is the maximal matching load threshold that is specified by users. When the value of *maxload* is smaller, the number of virtual frequent itemsets is larger.

3.2 Peer Initialization

In order to make full use of nodes and improve scalability of the system, the DHT-based overlay network is applied in DMPSS. As previously described, DMPSS can be deployed on any DHT-based overlay network, such as Chord, CAN, Pastry, etc. In DMPSS, after a peer joins in, it is initialized according to the DHT network that the user selects, and then it downloads frequent itemsets and their supports, as well as virtual frequent itemsets from existing peers.

3.3 Subscription Installation

A subscription is installed in terms of an attribute set which is a frequent itemset or a virtual frequent itemset, as shown in Algorithm 1. The attribute set is obtained as follows. The subscriber extracts the attribute set $Attr_{sub}$ from *sub*, and compares it with frequent itemsets. If $Attr_{sub}$ doesn't contain any frequent itemset, then one virtual frequent itemset is selected randomly. If $Attr_{sub}$ contains one frequent itemset, then this frequent itemset is the selected attribute set. If $Attr_{sub}$ contains more than one frequent itemsets, then the frequent itemset with the smallest support is selected. After the attribute set is specified, a string is obtained by bunching the names of attributes in the attribute set lexicographically. Then a key K is produced by hashing the string. Finally, the subscription is delivered to the related node in the overlay network. When more than one frequent itemsets are contained in $Attr_{sub}$, selecting the one with the smallest support can keep matching load balancing among RPs.

In DMPSS, each subscriber records all the subscriptions it sends and their corresponding keys. When a subscriber wants to cancel a subscription, it sends a cancel message to the node according to the key in recorder. Moreover, in order to avoid losing subscription information due to node departure, we suggest backing up subscriptions on neighbor nodes in the overlay network.

3.4 Event Publication

An event is published in terms of an attribute set list, as shown in Algorithm 2. The attribute set list is obtained as follows. The publisher extracts the attribute set $Attr_e$ from *e*, and checks if there are frequent itemsets are contained in $Attr_e$. All the frequent itemsets that are contained in $Attr_e$ are added to the attribute set list. Besides, all the virtual frequent itemsets are added to the attribute set list no matter whether they are contained in $Attr_e$. Assuming there are N attribute sets in the list,

---

Algorithm 1. Subscription Installation

Require:

   sub //the subscription
   vector < FrequentItemset > FIS[0...*n* − 1] //the set of frequent itemsets
   vector < FrequentItemset > VFIS[0...*m* − 1] //the set of virtual frequent itemsets
   extract(*se*) //returns attribute names in *se*
   IFContain(*A,F*) //returns TRUE if A contains F, FALSE otherwise
   LexiOrder(*Attr*) //returns a string by bunching the names of attributes in the attribute set
   lexicographically
   send(*message,key*) //deliver message to the node whose ID closet or immediately succeeds to key
1: set *AttrSet = NULL*;
2: set *support* = 1;
3: set *num* = 0;
4: set $Attr_{sub} = extract(sub)$;
5: **for** each $i \in [0, n-1]$ **do**
6:      **if** $IFContain(Attr_{sub}, FIS[i].itemset) == true$ **then**
7:          *num*++;
8:          **if** $FIS[i].support < support$ **then**
9:              $AttrSet = FIS[i].itemset$;
10:             $support = FIS[i].support$;
11:         **end if**
12:     **end if**
13: **end for**
14: **if** *num* == 0 **then**
15:     $i = RandomInteger(0, m-1)$;
16:     $AttrSet = VFIS[i].itemset$;
17: **end if**
18: $key = hash(LexiOrder(AttrSet))$;
19: $send(sub, key)$;

N strings will be generated by bunching the names of attributes in each attribute set lexicographically. Finally, the event is sent to the *N* related nodes in the overlay network for matching.

The event publication policy ensures that all subscriptions that match with an event are found. Moreover, because all subscriptions that don't contain frequent itemset are stored on no more than *M* RPs, the number of events that are sent for publication is reduced dramatically.

3.5   Subscriptions and Events Matching

When a RP receives an event *e*, it matches *e* with all the subscriptions stored on it and finds the matched subscribers. In DMPSS, subscriptions that contain the same frequent itemset can emerge strong covering relations, so covering-based matching algorithms are suggested to be used to improve matching efficiency.

Algorithm 2. Event Publication

Require:

    e //the event

1: set $Attr_e = extract(e)$;
2: **for** each $i \in [0, n-1]$ **do**
3:    **if** $IFContain(Attr_e, FIS[i].itemset) == true$ **then**
4:      $key = hash(LexiOrder(FIS[i].itemset))$;
5:      $send(sub, key)$;
6:    **end if**
7: **end for**
8: **for** each $i \in [0, m-1]$ **do**
9:    $key = hash(LexiOrder(VFIS[i].itemset))$;
10:   $send(sub, key)$;
11: **end for**

3.6 Event Delivery

RPs deliver the event e to all the matched subscribers after matching processes. In DMPSS, we use point-to-point communication to reduce cost on overlay routing and transmission. When a subscriber installs a subscription, it also sends its IP address to the RP. After the matching process finishes, the RP counts the number of matched subscribers, e.g. l, and collects their IP addresses. Then it calculates a transmission degree d as follows:

$$d = \frac{W}{S_e}$$

where W is the current available upload bandwidth, and S_e is the size of event e. If $l \leq d$, the RP sends event e to all the matched subscribers through direct point-to-point communication. If $l > d$, the RP divides all the matched subscribers into d groups, and randomly selects one node from each group as the group leader. Then it sends e and the IP list of the group to each leader. When a leader receives such a message, it has two choices. One of them is sending e to all the nodes in the IP list through direct point-to-point communication. The other is further dividing these nodes into groups and delivering e and the corresponding IP list to each sub-leader. A leader makes its choice also according to its available upload bandwidth.

3.7 Load Balancing Strategy

As mentioned above, the load on a node includes overheads for matching and over-heads for message transmission. In DMPSS, load balancing performance for message transmission is mainly affected by the routing policy of the DHT network, so we only provide load balancing strategy for matching.

The non-uniform distribution of matching load on nodes results from the skewed dataset of the real world. For example, in DMPSS, RPs related to frequent itemsets with higher supports will store more subscriptions, and their matching costs may be higher. Besides, when a RP receives much more events than other RPs, its matching costs may also be higher. In this paper, we relieve stresses of overloaded RPs by distributing matching overheads over more nodes.

When the RP related to the frequent itemsem FI is overloaded, its matching overheads are distributed to multiple nodes, and the number of newly increased nodes is decided by the number of attribute in FI. The detailed policy is presented as follows. If FI is selected as the attribute set for subscription installation, the subscriber ranks attribute names in FI in different order, and A_n^n keys will be obtained, where n is the number of attribute in FI. Then one key is randomly selected and the subscription will be sent to the corresponding node in the overlay network. For example, when the RP related to $\{A_1, A_2, A_3\}$ is overloaded, 6 keys can be obtained by hashing string $A_1A_2A_3$, $A_1A_3A_2$, $A_2A_1A_3$, $A_2A_3A_1$, $A_3A_1A_2$, and $A_3A_2A_1$. As a result, subscriptions that used to be stored on one RP will be distributed over 6 nodes, and the matching overheads will be reduced on each node. However, in order to realize exhausting matching, events that contain FI should be sent to all the newly increased nodes for matching. So the load balancing strategy may increase the event publication cost slightly.

4   Performance Evaluation

In this section, we evaluate the performance of DMPSS by comparing it with Ferry and Eferry. A set of metrics are used to evaluate performances of the three systems, including the total matching time on each RP, data volume of forwarded messages on each node, and latency of the event. As mentioned above, overhead for matching is affected by three factors: the number of subscriptions stored on the RP, the number of events the RP receives and the matching complexity. For evaluating the matching load comprehensively, we calculate the total matching time that is consumed on each RP during the event publication process. Data volume of forwarded messages indicates overhead for message transmission, and it also reflects the bandwidth consumption in the network. Latency is one of the most important performance metric for publish/subscribe systems, and it is denoted by the time duration from event publication to subscriber reception.

4.1   Experimental Setup

We built DMPSS, Ferry, and Eferry on OverSim [20] platform, which has implemented many DHT protocols. In order to realize equal comparison, Chord, which was used in Ferry and Eferry, is applied in DMPSS. Moreover, the schema used in our simulation is also the stock quotes model used in Ferry and Eferry. The definition of the schema is as follows:

$$S = \begin{cases} [Date: string, 2/Jan/98, 31/Dec/02], \\ [Symbol: string, aaa, zzzz], \\ [Open: float, 0, 500], \\ [Close: float, 0, 500], \\ [High: float, 0, 500], \\ [Low: float, 0, 500], \\ [Volume: integer, 0, 310000000]. \end{cases}$$

Symbol is the stock name. *Open* and *Close* are the opening and closing prices for a stock on a given day. *High* and *Low* are the highest and lowest prices for the stock on that day. *Volume* is the total amount of trade in the stock on that day.

We generated subscriptions by using six template subscriptions with different probabilities. The six templates are $T_1 = \{(Symbol = P_1) \wedge (P_2 \leq Open \leq P_3)\}$ with probability 20 percent, $T_2 = \{(Symbol = P_1) \wedge (Low \leq P_2)\}$ with probability 25 percent, $T_3 = \{(Symbol = P_1) \wedge (High \geq P_2)\}$ with probability 30 percent, $T_4 = \{(Symbol = P_1) \wedge (Volume \geq P_2)\}$ with probability 10 percent, $T_5 = \{(Volume \geq P1)\}$ with probability 5 percent, and $T_6 = \{(Date \geq P_1)(50\%) \wedge (Symbol = P_2)(50\%) \wedge (P_3 \leq Open \leq P_4)(50\%) \wedge (P_5 \leq Close \leq P_6)(50\%) \wedge (High \geq P_7)(50\%) \wedge (Low \leq P_8)(50\%) \wedge (Volume \geq P_9)(50\%)\}$ with probability 10 percent. In T_6, each predicate appears in the subscription with probability 50 percent. We used T_6 to generate irregular subscriptions.

Events in DMPSS were generated by seven template events with different probabilities. The seven templates are $E_1 = \{Date = P_1, Symbol = P_2, Open = P_3, Close = P_4\}$ with probability 10 percent, $E_2 = \{Date = P_1, Symbol = P_2, High = P_3, Low = P_4\}$ with probability 5 percent, $E_3 = \{Date = P_1, Symbol = P_2, Volume = P_3\}$ with probability 10 percent, $E_4 = \{Date = P_1, Symbol = P_2, Open = P_3, Close = P_4, High = P_5, Low = P_6\}$ with probability 5 percent, $E_5 = \{Date = P_1, Symbol = P_2, Open = P_3, Close = P_4, Volume = P_5\}$ with probability 30 percent, $E_6 = \{Date = P_1, Symbol = P_2, High = P_3, Low = P_4, Volume = P_5\}$ with probability 20 percent, and $E_7 = \{Date = P_1, Symbol = P_2, Open = P_3, Close = P_4, High = P_5, Low = P_6, Volume = P_7\}$ with probability 20 percent.

We set the minimum support threshold *minsup* to 10%. According to the six template subscriptions and their probabilities, we obtained 4 frequent itemsets: $\{Symbol, Open\}$ (20%), $\{Symbol, Low\}$ (25%), $\{Symbol, High\}$ (30%), and $\{Symb-ol, Volume\}$ (10%). The maximal matching load threshold *maxload* was set to 5%. According to probabilities of the template subscriptions, S_{NF} would larger than 85%, as a result, the number of virtual frequent itemsets was set to 3. In the simulation, the three virtual frequent itemsets were $\{Open, Close\}$, $\{High, Low\}$, and $\{Date, Volume\}$ respectively.

There were 1024 nodes in the simulation network. The number of subscriptions and events were 10240 and 102400, respectively.

4.2 Experimental Results

4.2.1 Overhead for Matching

We first compare the numbers of stored subscriptions on RPs. In DMPSS and Ferry, the numbers of RPs are both seven. Because the number of RPs in Eferry is large (more than one hundred), we rank the RPs by the number of stored subscriptions, and select the first 10 RPs for comparison. As shown in Fig. 1 and Table 1, DMPSS realizes better even storage distribution than Ferry-Pred and Eferry, and Ferry-Rnd has the best storage load performance. Besides, from Fig. 1 and Table 1 we can see that although Eferry increases the number of RPs, most of subscriptions are stored on a small part of nodes. As a result, the rest of RPs, which consume a large number of messages for matching, only store a small amount of subscriptions. However, in DMPSS, subscriptions that don't contain frequent itemsets are stored on only three RPs, and the event publication cost can be reduced dramatically.

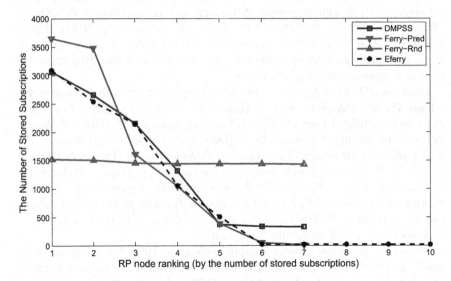

Fig. 1 The number of stored subscriptions on each RP

Table 1 The number of stored subscriptions on each RP

	RP_1	RP_2	RP_3	RP_4	RP_5	RP_6	RP_7	RP_8	RP_9	RP_10
DMPSS	3067	2655	2152	1320	376	339	331			
Ferry-Pred	3650	3480	1610	1050	390	50	10			
Ferry-Rnd	1523	1508	1456	1443	1442	1438	1430			
Eferry	3096	2539	2141	1057	511	23	21	20	19	17
DMPSS+LB	3148	2614	1295	1069	1019	374	372	349		

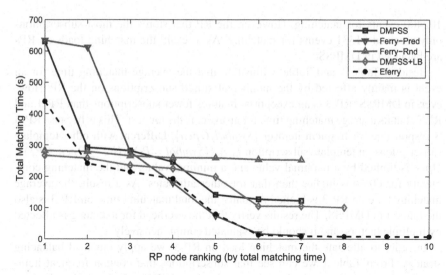

Fig. 2 The total matching time on each RP

Table 2 The number of received events on each RP

	RP_1	RP_2	RP_3	RP_4	RP_5	RP_6	RP_7	RP_8	RP_9	RP_10
DMPSS	51331	51331	66655	81760	102400	102400	102400			
Ferry-Pred	102400	102400	102400	102400	102400	102400	102400			
Ferry-Rnd	102400	102400	102400	102400	102400	102400	102400			
Eferry	51412	51412	66616	81909	79933	77401	87400	144019	72209	41574
DMPSS+LB	51413	51413	81839	66503	66503	102400	102400	102400		

Table 3 The average matching time for an event on each RP(ms)

	RP_1	RP_2	RP_3	RP_4	RP_5	RP_6	RP_7	RP_8	RP_9	RP_10
DMPSS	5.4916	4.7900	9.4579	3.5720	1.3814	1.1952	1.2437			
Ferry-Pred	6.1936	5.9771	2.7091	1.7438	0.7186	0.1600	0.0828			
Ferry-Rnd	2.7571	2.7232	2.5300	2.5285	2.4634	2.5908	2.4767			
Eferry	4.6944	4.1609	6.5946	2.3502	0.9865	0.1004	0.0911	0.0900	0.1034	0.1450
DMPSS+LB	4.5712	3.8518	2.7472	4.0108	3.8806	1.0270	1.0218	0.9815		

As mentioned above, the numbers of stored subscriptions can't evaluate the overhead for matching comprehensively. So we calculate the total matching time that is consumed on each RP during the event publication process, and the result is shown in Fig. 2. Similar to results of Fig. 1, DMPSS realizes better even overhead for matching than Ferry-Pred and Eferry, and Ferry-Rnd has the best matching load performance. We record the numbers of received events and the average matching time for an event on RPs, and list them in Table 2 and Table 3. As shown in the tables, in DMPSS, RP_5, RP_6, and RP_7 store less than 400 subscriptions, and they receive

102400 events for matching. However, the RP that stores the most subscriptions only received 51331 events for matching. As a result, the matching loads on RPs are balanced in DMPSS.

Besides, Table 1 and Table 3 illustrate that the average matching time for an event is mainly affected by the number of stored subscriptions on the RP. However, in DMPSS, RP_3 is an exception. It stores fewer subscriptions than RP_1 and RP_2, but its average matching time for an event is the longest. This is because RP_3 is responsible for frequent itemset $\{Symbol, Open\}$. Different with other template subscriptions, in template subscription $T_1 = \{(Symbol = P_1) \wedge (P2 \leq Open \leq P3)\}$, $Open$ is limited by a maximal value and a minimal value, and the matching complexity for $Open$ is higher than that for other attributes. As a result, the average matching time on RP_3 is higher. Actually, the total matching time on RP_3 is also the longest in DMPSS. The results verify that the overhead for matching is affected by multiple factors, and it should be evaluated comprehensively.

In order to alleviate the matching load on RP_3, we apply our load balancing strategy. From Table 1 we can see that subscriptions that contain frequent itemset $\{Symbol, Open\}$ are distributed over one more RP. Fig. 2 shows that the load balancing performance is improved obviously when the load balancing strategy is applied.

4.2.2 Overhead for Message Transmission

We calculate the data volume of forwarded messages on each node to evaluate the overhead for message transmission and the bandwidth consumption in the network. From Fig. 3 we can see that DMPSS reduces the data volume of forwarded messages on each node obviously. This is because the number of events that are published for matching is reduced in DMPSS. Besides, the point-to-point communication we used for event delivery reduces cost on overlay routing and transmission, and the overhead for message transmission is reduced further. When our load balancing strategy is applied, more event messages are published for matching, and the data volume of forwarded messages increases slightly, as shown in Fig. 2.

4.2.3 Latency

Fig. 4 plots the cumulative distribution functions of event latency. We can see that DMPSS decreases the time duration from event publication to subscriber reception dramatically. The reduction in latency mainly results from two reasons. First, the number of events that are published for matching is reduced, and the throughput in the network is reduced accordingly. Second, the point-to-point communication in DMPSS avoids time for overlay routing and transmission, and the latency is reduced correspondingly. As described in [6], compared to Ferry-Rnd, Ferry-Pred can reduce hops, latency and overhead, so Ferry-Pred has better latency performance than Ferry-Rnd. However, it is still worse than DMPSS. Eferry combines the event publication into event delivery. This strategy can reduce the number of messages in

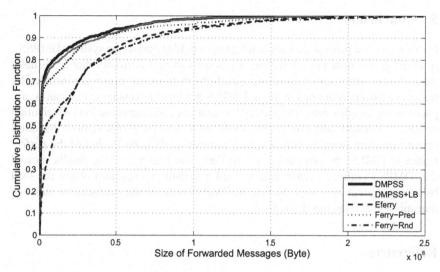

Fig. 3 The cumulative distribution functions for size of forwarded messages

the system, but the latency is increased because of the matching process on every node in the embedded tree. As a result, the latency in Eferry is the longest.

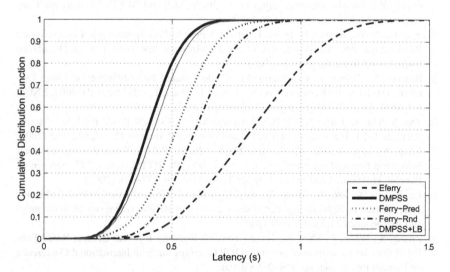

Fig. 4 The cumulative distribution functions for latency

5 Conclusion

In this paper, we proposed a data mining based publish/subscribe system (DMPSS). In DMPSS, the data mining technology is used to find attributes that are usually subscribed together, e.g. frequent itemset. Subscriptions and events are installed by frequent itemsets contained in them. DMPSS has three advantages: 1) it can be applied on any overlay network easily; 2) it achieves even matching load distribution on RPs; 3) it reduces the event publication cost by grouping subscriptions that don't contain frequent itemsets onto a certain number of RPs. We evaluated the performance of DMPSS by comparing it with Ferry and Eferry, and the results showed that DMPSS realized even matching load distribution, especially when the load balancing strategy was used. Moreover, DMPSS reduced the overhead for message transmission and latency dramatically.

References

1. Shen, H.: Content-Based Publish/Subscribe Systems. In: Handbook of Peer-to-Peer Networking, pp. 1333–1366 (2010)
2. Rahimian, F., Girdzijauskas, S., Payberah, A.H., Haridi, S.: Subscription Awareness Meets Rendezvous Routing. In: The Fourth International Conference on Advances in P2P Systems, pp. 1–10 (2012)
3. Gupta, A., Sahin, O.D., Agrawal, D., et al.: Meghdoot: Content-Based Publish/Subscribe over P2P Networks. In: Proceedings of the 5th ACM/IFIP/USENIX International Conference on Middleware, pp. 254–273 (2004)
4. Terpstra, W.W., Behnel, S., Fiege, L., et al.: A Peer-to-Peer Approach to Content-Based Publish/Subscribe. In: Proceeding of the 2nd International Workshop on Distributed Event-Based Systems, pp. 1–8 (2003)
5. Bianchi, S., Felber, P., Gradinariu, M.: Content-Based Publish/Subscribe Using Distributed R-Trees. In: Kermarrec, A.-M., Bougé, L., Priol, T. (eds.) Euro-Par 2007. LNCS, vol. 4641, pp. 537–548. Springer, Heidelberg (2007)
6. Zhu, Y., Hu, Y.: Ferry: A P2P-Based Architecture for Content-Based Publish/Subscribe Services. IEEE Transactions on Parallel and Distributed Systems 18(15), 672–685 (2007)
7. Yang, X., Zhu, Y., Hu, Y.: Scalable Content-Based Publish/Subscribe Services over Structured Peer-to-Peer Networks. In: 15th EUROMICRO International Conference on Parallel, Distributed and Network-Based Processing, pp. 171–178 (2007)
8. Lu, P., Liu, X., Lin, X., Wang, B.: Key Algorithm in Content-based Publish/Subscribe system based on Subscription Partitioning. Journal of Beijing University of Aeronautics and Astronautics 32(8), 992–997 (2006)
9. Zhang, S., Wang, J., Shen, R., Xu, J.: Towards Building Efficient Content-Based Publish/Subscribe Systems over Structured P2P Overlays. In: 39th International Conference on Parallel Processing, pp. 258–266 (2010)
10. Rahimian, F., Girdzijauskas, S., Payberah, A.H., Haridi, S.: Subscription awareness meets rendezvous routing. In: The Fourth International Conference on Advances in P2P Systems, pp. 1–10 (2012)
11. Stoica, I., Morris, R., Karger, D., Kaashoek, M., Balakrishnan, H.: Chord: A Scalable Peer-to-Peer Lookup Service for Internet Application. In: Proc. ACM SIGCOMM, pp. 149–160 (2001)

12. Ratnasamy, S., Francis, P., Handley, M., Karp, R., Shenker, S.: A Scalable Content-Addressable Network. In: Proc. of ACM SIGCOMM 2001, pp. 329–350 (2001)
13. Zhao, B., Kubiatowicz, J., Joseph, A.: Tapestry: An Infrastructure for Fault-Tolerant Wide-Area Location and Routing. Technical Report UCB/CSD-01-1141, Computer Science Division, UC Berkeley (2001)
14. Rowstron, A., Druschel, P.: Pastry: Scalable, Decentralized Object Location and Routing for Large-Scale Peer-to-Peer Systems. In: Proc. of the 18the IFIP/ACM International Conference on Distributed Systems Platforms (Middleware) (2001)
15. Rowstron, A., Kermarrec, A.-M., Druschel, P.: SCRIBE: The design of a large-scale event notification infrastructure. In: Crowcroft, J., Hofmann, M. (eds.) NGC 2001. LNCS, vol. 2233, p. 30. Springer, Heidelberg (2001)
16. Zhuang, S.Q., Zhao, B.Y., Joseph, A.D., Katz, R.H., Kubiatowicz, J.: Bayeux: An Architecture for Scalable and Fault-Tolerant Wide-Area Data Dissemination. In: Proc. 11th Intl Workshop Network and Operating System Support for Digital Audio and Video (NOSSDAV 2001) (2001)
17. Han, J., Cheng, H., Xin, D., Yan, X.: Frequent Pattern Mining: Current Status and Future Directions. Data Min. Knowl. Disc., 55–86 (2007)
18. Han, J., Pei, J., Yin, Y.: Mining Frequent Patterns without Candidate Generation. In: SIGMOD 2000, pp. 1–12 (2000)
19. Agrawal, R., Srikant, R.: Fast Algorithms for Mining Association rules. In: VLDB 1994, pp. 487–505 (1994)
20. Baumgart, I., Heep, B., Krause, S.: OverSim: A Flexible Overlay Network Simulation Framework. In: 10th IEEE Global Internet Symposium (GI 2007) in Conjunction with IEEE INFOCOM 2007, pp. 79–84 (2007)

Secure Identity Management System for Federated Cloud Environment

Umme Habiba, Rahat Masood, and Muhammad Awais Shibli

Abstract. Federated Identity Management (FIM) systems are well-known for achieving reliable and effective collaboration among various organizations. Despite numerous benefits, these systems have certain critical weaknesses such as lack of security and privacy while disseminating identity credentials (Personally Identifiable Information (PII)) across multiple federated Cloud environments. In addition to this, FIM systems have limitations in terms of interoperability and lack compliance to international standards, since most of the systems are reliant on proprietary protocols for the exchange of identity information. In this regard, we propose a secure identity management system for federated Cloud environments that not only ensures the secure management of identity credentials, but preserves the privacy of Cloud Service Consumers (CSC) also. Furthermore, implementation of the proposed system involves state-of-the-art international standards *(SCIM, SAML, REST and XACML)* to ensure secure, quick and easy sharing & management of identity credentials in to, out of and around the Cloud. Further, we have performed rigorous evaluation of the proposed system using standard evaluation tools such as Scyther and JUnit. The results of our evaluation certify that the presented work ensures the desired features and level of security as expected from a secure identity management system for federated Cloud environment.

Keywords: (Identity management systems, cross-domain identity management, Access right delegation, identity synchronization, Cloud computing).

Umme Habiba · Rahat Masood · Muhammad Awais Shibli
School of Electrical Engineering & Computer Science,
National University of Science & Technology,
Islamabad, Pakistan
e-mail: {11msccsuhabiba,10msccsmmasood,awais.shibli}@seecs.edu.pk

1 Introduction

Considering the widespread benefits and support for state-of-the-art business requirements, such as ease of service delivery and collaborative resources, Cloud computing is actively adopted by the small and medium size business organizations. It facilitates the federation and collaboration among independent business organizations while acquiring the services and resources from disparate Cloud environments. Identity management service being the foundation for all the other services is ideal to be out-sourced, since this is the first and foremost service that is invoked by the collaborating organizations. Similarly, in federated Cloud computing environments, where an agreement is made among the group of trusted *Cloud Service Providers* (CSPs) to share their services and resources in demand spikes; identity credentials are required to be exchanged and shared. For such scenarios, FIM is the common deployment model, where CSPs with mutual trust relationship share the identity information of their subscribers with each other on demand or as per the requirement [1, 2]. With the introduction of FIM systems, Cloud subscribers are able to use the same identity credentials for gaining access to the set of shared Cloud resources. FIM brings in economic benefits along with the convenience to the participating organizations and its network subscribers.

However, federated Cloud environment also raises many security and privacy concerns regarding the handling and sharing of sensitive identity information [3], such as *Who has the access to identity information?, How to maintain identical and updated user information across multiple CSPs?, How to ensure standard security and privacy procedures across multiple CSPs?* and *How much control does the subscriber has over his information?*. In order to answer these questions, several security based FIM systems have been designed and implemented. But, to the best of our knowledge, none of those systems offer a holistic solution covering self-service, privacy and real-time synchronization features for Cloud computing environments [4]. In addition to this, most of the existing software applications are bundled with proprietary identity management services and authentication mechanisms. As a result, interoperability has become an increasingly significant challenge in the federated Cloud computing environment.

We have designed and implemented a *Secure Identity Management System for Federated Cloud environments* that ensures seamless integration and utilization of identity credentials. In addition to the basic identity management features, such as provisioning, de-provisioning and user account management; we intend to provide advanced security features as well. Those advanced security features include *access right delegation, real-time synchronization and self-service* in cross-domain Cloud computing scenarios. Further, for the implementation of these advanced features, we have used state-of-the-art international standards *(such as SCIM, SAML, REST and XACML)* that guarantee secure, quick and easy sharing & management of identity credentials in to, out of and around the Cloud. Finally, we present E-Healthcare system as a case-study to explicate the work-flow and various use-case scenarios of the implemented system. The rationale behind choosing E-healthcare systems as a case-study lies with their significance for dealing with sensitive and critical identity and user information.

This book chapter is organized as follows: Section 2 presents the related work covering IDMS from an inter-Cloud perspective. Components of the proposed Secure IDMS for federated Cloud environment along with a complete work-flow is discussed in Section 3. Section 4 presents the use-case scenarios of the proposed system. However, details regarding the evaluation tools and techniques with corresponding results are presented in Section 5. Finally, in Section 6 we conclude this chapter by highlighting our contribution and future research directions.

2 Related Work

Identity management is among the top security challenges that organizations face while moving their critical identity data at Cloud. Cloud identity management systems may exist in different flavors, however, Federated IDMS is the common deployment model used for inter-Cloud environments. Several IDMSs [5, 6, 7] have been proposed in the past few years; however, here we explain only a few well-known Cloud based FIM systems that are closely related to the proposed secure Identity Management System for federated Cloud environment. Celesti et al. in [8] have proposed an Inter-Cloud Identity Management Infrastructure (ICIMI) where Home Cloud forwards the federation request to Foreign Cloud with the aim of expanding its virtualization infrastructure. Similarly, Yan et al. in [9] have also discussed various issues regarding web-service security and proposed a Federated Identity Management (FIM) system for Cloud along with Hierarchical Identity-Based-Cryptography (HIBC). Security APIs for My Private Cloud is presented in [10] that discusses a set of three security APIs. These APIs are designed for allowing the users to delegate their access rights to anyone at any time and offers federated access rights to Cloud resources.

We have analyzed various identity management systems [8, 9, 11] and our study reveals that none of the existing systems heuristically cover all the required features including privacy, user-centricity, real-time synchronization, provisioning, de-provisioning, interoperability and access right delegation; which are much needed for federated Cloud IDMSs. Further, each of the analyzed system has its own certain weaknesses regarding interoperability, implementation and deployment. For instance, *Security APIs for my Private Cloud* [10] offers PHP based implementation for the proposed solution, clearly limiting its utility to PHP based applications only, hence lacks interoperability. In addition to this, it offers no security mechanism to protect its authorization database and access right delegation tokens from illegal modification and forgery. Therefore, considering the prospective adoption of FIM systems, we propose a secure identity management system for federated Cloud environment that ensures interoperability, access right delegation, real-time synchronization along with communication level security and privacy. All of the aforementioned features collectively provide the desired level of security, privacy and interoperability.

3 Proposed Secure IDMS for Federated Cloud Environment

In this section, we present the design and architecture of *Secure Identity Management System for Federated Cloud Environment*. We have enhanced the conventional IDMS by proposing and designing the secure Identity Management System for federated Cloud that addresses the security and interoperability concerns of this emerging paradigm. We have based our system on well-known industry standards such as XACML, REST, SAML and SCIM schema. The layered architecture proposed for Secure Identity Management System for Federated Cloud Environment along with the related components are presented in Figure 1. Brief description of each system module is provided in the following subsections.

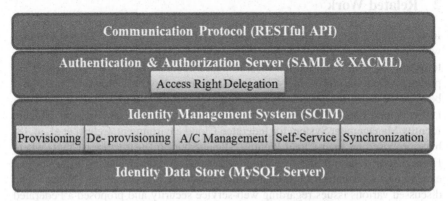

Fig. 1 Architecture of Secure IDSM for Federated Cloud Environment

1. **Provisioning:** This module is responsible for the on-boarding or creation of user accounts using SCIM schema and REST. In the proposed system, after receiving the provisioning request from CSC, Cloud administrators collects and verifies all the required information and stores it in its identity data store. They further associate the CSC with some specific role according to the collected information, so that CSC's future requests for Cloud services and resources are evaluated in the light of privileges assigned to that specific role. Cloud administrator then dispatches the CSC's information related to his verification and authorization to the respective authentication and authorization servers. Listing 1 provides a non-normative example of SCIM user representation in JSON format.

2. **De-Provisioning:** or Account Deactivation module deals with the real-time and synchronized deletion or suspension of user accounts. Upon receiving the deactivation request, Cloud administrator issues an access revoke request to the authorization server for immediate deactivation or suspension of CSC's access rights from all the Cloud services and resources. However, in case of any ongoing activity or transaction, CSCs will be prompted with an error message, stating "Your access rights have been revoked" followed by an immediate termination of their active sessions.

Listing 1 SCIM User Provisioning

```
{"schemas":[urn:scim:schemas:core:1.0,
    "urn:scim:schemas:extension:enterprise:1.0"],
userName:"Umme",
externalId:"Uhabiba",
name:{"formatted":"Umme
    Habiba","givenName":"Umme","familyName":"Ali"},
password:"123@sadf@@wsd",
id:"ghf_1245",
roles:[{"value":"Patient"}],
locale:"Islamabad",
preferredLangauge:"English",
addresses:[{"streetAddress":"Sector
    H-12","primary":true}],
emails:[{"value":"11msccsuhabiba@seecs.edu.pk"}],
phoneNumbers:[{:[{"value":"92334532589"}]}}]
```

3. **User Account(A/C) Management:** User account management module is responsible for the management of changes in user account throughout their identity's lifetime. Cloud subscribers may subscribe for new Cloud services or resources; such activities are reliably adjusted for the respective user across multiple Cloud servers. For instance, if change in user's access privileges or attribute values is encountered then that account change is made accordingly and timely in a synchronized manner.

4. **Authentication:** Provisioning module forwards the information regarding user authentication such as user name, user-ID, password etc. to the authentication server. Our authentication server implements SSO authentication using SAML 2.0 SSO profile, which allows the CSC to access multiple distinct Cloud resources with just one time authentication within a single session. Furthermore, security and privacy of identity credentials during the communication is ensured through encrypted SAML authentication request and response. Sample SAML authentication query request and response are presented in Listing 2 and Listing 3 for better understanding.

5. **Authorization:** This module is responsible for ensuring legitimate access to Cloud services and resources. We have implemented *Role-Based Access Control model (RBAC)* using XACML, that creates and enforces the access control policies for the subjects (CSCs) based on their roles assigned by the provisioning module.

Listing 2 SAML Authentication Request

```
<samlp:AuthnQuery
    xmlns:samlp="urn:oasis:names:tc:SAML:2.0:protocol"
    ID="AuthnQuery1">
 <saml:Subject
     xmlns:saml="urn:oasis:names:tc:SAML:2.0:assertion">
  <saml:NameID>"Umme"</saml:NameID>
 </saml:Subject>
 <samlp:RequestedAuthnContext>
 <saml:AuthnContextClassRef
     xmlns:saml="urn:oasis:names:tc:SAML:2.0:assertion">
  urn:oasis:names:tc:SAML:2.0:ac:classes:
     "PasswordProtectedTransport"
 </samlp:RequestedAuthnContext>
</samlp:AuthnQuery>
```

Listing 3 SAML Authentication Response

```
<samlp:Response
    xmlns:samlp="urn:oasis:names:tc:SAML:2.0:protocol"
    InResponseTo="AuthnQuery1">
 <samlp:Status>
  <samlp:StatusCode
      Value="urn:oasis:names:tc:SAML:2.0:status:Success"/>
 </samlp:Status>
   <saml:Subject>
       <saml:NameID>"Umme"</saml:NameID></saml:Subject>
   <saml:AuthnStatement> <saml:AuthnContext>
   <saml:AuthnContextClassRef>urn:oasis:names:tc:SAML:2.0:
   ac:classes:"PasswordProtectedTransport"
   </saml:AuthnContextClassRef>
  </saml:Assertion>
</samlp:Response>
```

6. **Access Right Delegation (ARD):** This module deals with the delegation of access rights among the subscribers of local as well as different but trusted Cloud domains. Considering E-Healthcare system as a case study, proposed system dynamically generates new access control policies by gathering the information related to *subject* (DoctorA), *resource* (Patient's health record), *action* (view) and *environment* (condition- such as date/time/IP) and passing it to the PAP for policy generation. Sample XACML based policy that allows Doctor A to view Patient B's record if he is accessing the resource through IP 192.168.0.1, is presented in Listing 4.

Listing 4 XACML based ARD Policy

```
<Policy PolicyId="Policy:id:1"
   RuleCombiningAlgId="urn:oasis:names:tc:xacml:1.1:
rule-combining-algorithm:ordered-permit-overrides">
   <Target/>
 <Rule RuleId="Rule:id:1" Effect="Permit">
 <Description>"This Policy applies to Doctor A, who can
   view Patient B's records using
     198.162.0.1"</Description>
 <Target>
  <Subjects>     <Subject>
  <SubjectMatch MatchId="urn:oasis:names:tc:xacml:1.0:
   function:string-equal"> <AttributeValue
      DataType="http://www.w3.org/2001/XMLSchema#string">
      Doctor A </AttributeValue>
   </SubjectMatch> </Subject> </Subjects>
  <Resources> <Resource>
  <ResourceMatch MatchId="urn:oasis:names:tc:xacml:1.0:
   function:anyURI-equal"> <AttributeValue
      DataType="http://www.w3.org/2001/XMLSchema#anyURI">
     Patient B </AttributeValue>
   </ResourceMatch> </Resource> </Resources>
  <Actions> <Action>
   <ActionMatch MatchId="urn:oasis:names:tc:xacml:1.0:
  function:string-equal"> <AttributeValue
      DataType="http://www.w3.org/2001/XMLSchema#string">
     View    </AttributeValue>
   </ActionMatch> </Action> </Actions>
 </Target>
 <Condition FunctionId="urn:oasis:names:tc:xacml:1.0:
  function:string-equal"> <AttributeValue
     DataType="http://www.w3.org/2001/XMLSchema#anyURI">
   192.168.0.1 </AttributeValue>
 </Condition> </Rule>
  <Rule RuleId="Rule:id:default" Effect="Deny">
</Policy>
```

All of the generated policies are then placed at a centralized and secure policy repository accessible to both CSPs. Our system does not include delegation chains, meaning any subject that has delegated (temporary) access rights is not allowed to further delegate the access rights, since they do not own the resource they are attempting to delegate. After successful policy generation, whenever, the PDP receives the resource access request from PEP, it evaluates the policy against the provided request and replies back to the PEP for the enforcement of its decision. Proposed system ensures the privacy of CSCs also, in the context of the same E-Healthcare system (Doctor and Patient), during the registration phase Cloud administrators asks the Patient that whether or not they allow their

doctor to delegate the view/access rights of their medical record to some other
doctor, and if they choose 'No' then no delegation will be performed at all.

7. **Synchronization:** This module provides real-time synchronization of CSC's
accounts across multiple Cloud servers. During the CSC's provisioning phase
at CSP1 Cloud administrators asks for CSC's consent regarding whether or not
they want to synchronize their account information on CSP2. Based on their
privacy preferences this module is invoked to perform the synchronization of
CSC's account information. The prime objective of credentials synchronization
is to ensure that each Cloud IDMS holds identical information. In an Inter-
Cloud environment, synchronization is critical to be ensured, otherwise in case
of fail-over, out-of-sync data causes many problems.

8. **Self-Service:** Self-service module allows the user to update his account information such as e-mail address, password, postal address etc. only through the
self-service module. In addition, this module is responsible for making the same
and consistent change across multiple Cloud servers to ensure synchronized and
consistent user information. It facilitates the CSCs by offering them with firm
control over the sharing of their sensitive identity credentials by taking their
consent prior to each credential exchange. However, sensitive information such
as role and other unique identification information that may effect system security is not allowed to be modified.

3.1 Workflow of Secure IDMS for Federated Cloud Environment

The basic work-flow of CSC1's registration at the Secure Identity Management System for Federated Cloud Environment is presented in Figure 2. A brief description
for each step is also presented below,

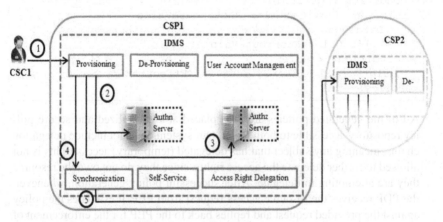

Fig. 2 Work flow of Secure IDSM for Federated Cloud

Step 1. CSC1 registers at CSP1 by providing all the required information to the Cloud administrators. Cloud administrator then invokes user provisioning service which is responsible for the creation of user account. Provisioning module collects necessary user information and stores it in its authoritative identity data store.

Step 2. Provisioning module forwards the information regarding user authentication to authentication server for the handling of forthcoming authentication requests.

Step 3. Similarly, information regarding user role and access rights is dispatched to the authorization server for the formulation and enforcement of access control policies.

Step 4. After successful user provisioning and distribution of information among severs, synchronization module will be invoked by the provisioning module depending upon the privacy preferences of CSC1.

Step 5. (Optional) In order to synchronize the authoritative identity data stores Synchronization module makes a connection with CSP2. As a result of this synchronization an identical account is created on CSP2 as well. The same process of provisioning will be performed on CSP2, following through Step 1 till Step 5 until all the Cloud servers hold identical user accounts.

4   Use Case Scenario

There are multiple use-cases of proposed secure IDMS for federated Cloud environment, such as user-account de-provisioning, user account synchronization, user carrying out self-service etc. However, here we discuss two possible use-cases of the proposed system by taking E-healthcare systems as a case-study. In use-case 01, we explain the process of access right delegation among the subscribers of distinct but trusted Cloud domains, whereas in use-case 02, we elaborate the steps that the CSC takes to use the delegated service in an inter-Cloud environment.

4.1   Access Right Delegation (ARD)- Use-Case 01

Use-case 01 depicts an access right delegation scenario, where CSC1 (Doctor A) is a subscriber of CSP1 and attempts to delegate his access rights on a Patient A's record to CSC2 (Doctor B) a subscriber of CSP2. In order to perform this delegation, Doctor B must be trusted for Doctor A and CSP1 also. As a result of this delegation, Doctor B will be able to access and view the medical record of Patient A on the behalf of Doctor A. A brief overview of the access right delegation use-case is shown in Figure 3; however, we elaborate each step below,

1. *Doctor A* forwards his identity credentials in the form of SAML request to the authentication server for verification.

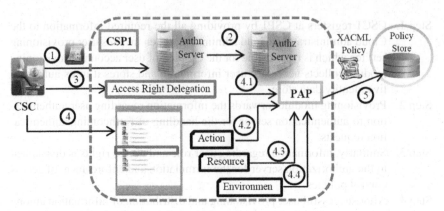

Fig. 3 Access Right Delegation

2. After successful authentication, user credentials are forwarded to the authorization server for the confirmation of his access rights. Request contains *<Au-
thzDecisionStatement>* comprising of information regarding *<Subject>, <Ac-
tion>, <Resource>, <Decision>* and an optional *<Evidence >* attribute.
3. In the next step, user attempts to delegate his access rights by initiating an access
right delegation request.
4. *Doctor A* is provided with a view where he specifies the required attributes for
policy generation

    4.1. List of doctors that are trusted for him are displayed and he chooses
    Doctor B from that list, so that he may access Patient A's records on the
    behalf of Doctor A.
    4.2. Action (such as view, edit) is then specified against the patient that the
    delegator *(Doctor A)* wants to delegate
    4.3. Similarly, list of patients is presented so that the *Doctor A* specifies the
    patient name whose record accession rights are to be delegated and from
    that list he chooses *Patient A*.
    4.4. Finally, delegator *(Doctor A)* specifies the duration, for which the dele-
    gatee *(Doctor B)* can access Patient A's records.

    All the required information is submitted to the PAP for the generation of access
    control policy. This information includes the identity of delegator (Doctor A),
    delegatee (Doctor B), time for which the application is delegated and informa-
    tion regarding the resource (Patient A) being delegated.

5. After successful generation of access right delegation token, XACML based
policy is generated and stored at a central policy repository that is accessible to
all the CSPs.

4.2 Usage of Delegated Access Rights by CSC - Use-Case 02

In use-case 02, we present a list of steps that the user might take to use a delegated service. In particular this use-case depicts the complete set of steps that will be performed after the execution of access right delegation explained in the previous sub-section, it further assumes that the CSC is authenticated and authorized user of Cloud. A brief overview of service usage is presented in Figure 4; however, we elaborate each step below,

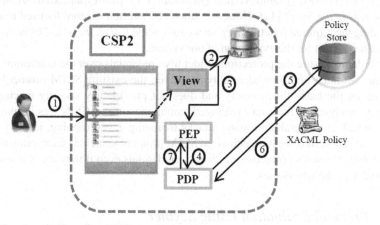

Fig. 4 Synchronization after ARD

1. Doctor B selects the record of Patient B.
2. Click on the view button result in a database query.
3. However, prior to record display, PEP captures the view record request, collects the required attributes and prepares the query for PDP server.
4. As a next step, PEP forwards the prepared query to the PDP server for evaluation.
5. PDP server, upon receiving the evaluation requests queries its Policy store.
6. Policy store responds back with the list of most related policies.
7. PDP server evaluates the request against the provided policy set and gives its decision (permit/deny). PEP then enforces the returned decision and either presents the medical record of Patient B or displays an error message specifying the details, accordingly.

5 Evaluation

Cloud Security Alliance (CSA) provides *Guidance for Identity & Access Management V2.1* [12], which offers useful practical guidelines and recommendations which may help software developers to design and analyze the functionality aspects of their Cloud based IDMSs. According to CSA, secure IDMS for federated Cloud environment must include real-time and synchronized user provisioning, de-provisioning, interoperability and communication level security. In addition to this, commercial identity management systems offered by well-known vendors such as *Hitachi ID Identity Manager* [13], *NetIQ's Identity Manager Standard 4 Edition* [14], *Microsoft's Enterprise Identity Synchronization Architecture* [15] and *McAfee Cloud Identity Manager* [16] among many others also provide documents describing recommended features for federated identity management systems and define guidelines for evaluation. Similarly, NIST [17] offers well-established aspects that could be used to evaluate the functionality of a system.These aspects include *Correctness and Effectiveness, Leading versus Lagging Indicators, Organizational Security Objectives, Qualitative and Quantitative Properties* and *Measurement of Large versus the Small* [17]. However, since our solution is more focused towards providing the required functionality, so we follow the *Correctness and Effectiveness Properties* aspect for the evaluation of our system.

In order to ensure the protection of identity credentials over the communication channel we must offer confidentiality. However, the existing SCIM protocol only focuses on the basic functionality and does not provide any security mechanism at all. Consequently, the data exchanged among the various CSPs was in plaintext form, which is vulnerable to many attacks including eavesdropping, identity theft and fraud. Therefore, we have enhanced SCIM protocol to include encryption mechanism which ensures protection of identity credentials even if they are disclosed or accessible to the adversaries.

5.1   Protocol Evaluation Using Scyther

Scyther is a well-known tool that is used for the formal analysis of security protocols under the perfect cryptography assumption, it verifies that all the cryptographic functions used in the system are perfect, for-instance it may validate that the attacker learns nothing from an encrypted message until and unless he has the decryption key. Therefore, we have used Scyther to confirm the effectiveness of enhanced SCIM protocol. Scyther has verified the claims of our protocol, like secrecy of identity credentials (in request and response) and the persistence of server aliveness is ensured during the identity exchange processes. It has verified that the security enhancement in SCIM protocol ensures protection against the intruder even if the network is completely or partially under his control. Scyther lists various predefined claim types such as *NiSynch, WeakAgree, Alive, NiAgree, Empty, Reachable* and *Secret* [18]. However, we have used only four of them (Secret (for confidentiality), Alive, WeakAgree and NiSynch) as per our requirement. The script of Scyther, stating steps of our protocol in the form of claims along with the verification results are presented in Listing 5 and Figure 5 respectively.

Correctness and Effectiveness Properties: Correctness property aims to ensure that the required mechanisms are properly implemented; however, effectiveness ascertains how well the various system components tie together and work in synergy [17]. Generally, functionality assessments using correctness and effectiveness are essentially performed through direct evaluation of actual system components. In the same way, we have designed various test-cases to confirm the effectiveness of implemented features and highlight the methods which have been used to achieve those features.

Listing 5 SCYTHER Script

```
#### Secure SCIM Protocol ####
usertype SessionKey, String;
protocol Secure-SCIM(CSP1,CSP2)}
{
 { Role CSP1 }
  {
    fresh identity-Credentials: String;
    fresh request: String;
    var response: String;
    var responsecode: String;
    var pre-sharedKey: SessionKey;
    send\_1(CSP1,CSP2,CSP1,{request}k(CSP1,CSP2));
    recv\_2(CSP2,CSP1,{request,response} k(CSP1,CSP2));
    send\_3(CSP1,CSP2,{response}k(CSP1,CSP2));
    recv\_4(CSP2,CSP1,{pre-sharedKey,responsecode,
            request} k(CSP1,CSP2));
    claim\_I1(CSP1,Alive);
    claim\_I2(CSP1,Nisynch);
    claim\_I3(CSP1,Weakagree);
    claim\_I4(CSP1,Secret, pre-sharedKey);
  }
 { Role CSP2 }
  {
    var identity-Credentials: String;
    var request: String;
    fresh response: String;
    fresh responsecode: String;
    fresh pre-sharedKey: SessionKey;
    recv\_1(CSP1,CSP2,CSP1,{request}k(CSP1,CSP2));
    send\_2(CSP2,CSP1,{request,response}k(CSP1,CSP2));
    recv\_3(CSP1,CSP2,{response}k(CSP1,CSP2));
    send\_4(CSP2,CSP1,{pre-sharedKey,responsecode,
            request} k(CSP1,CSP2));
    claim\_R1(CSP2,Alive);
    claim\_R2(CSP2,Nisynch);
    claim\_R3(CSP2,Weakagree);
    claim\_R4(CSP2,Secret, pre-sharedKey);
  }
}
```

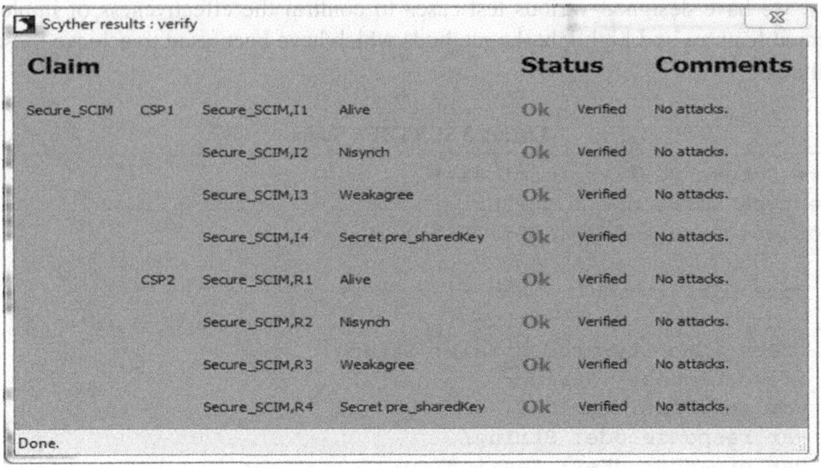

Fig. 5 Scyther Evaluation Output

5.2 Protocol Evaluation Using Test-Cases

In order to confirm the correctness and effectiveness of implemented functionality, we have designed and executed various test-cases using JUnit4. JUnit4 is a Java test framework that offers annotations based flexible modeling utility. We have performed a high test coverage of our code, which confirms that each of the implemented functionality works as intended. However, here we present the results of only two important unit test-cases in the form of test reports. Table 1, 2 provide further details:

Table 1 Test Case 01

Test Case Title	Self-service Test
Test Case ID	Test 01
Test Objective	To confirm the correctness and effectiveness of self-service feature.
Pre-Condition	PatientX must be logged in at CSP1
Post-Condition	PatientX shall be able to login at CSP2 as well
Procedure	**Assuming PatientX has opted for no synchronization at the time of his registration at CSP1** 1. Select **View User** option from the main menu 2. Checks *Synchronize* across CSP2 option. 3. Clicks on **Save** button and close the tab. **Switch tab: CSP2:** 4. Enter his authetication credentials. 5. Press Log-in
Expected Result	PatientX Logged in Successfully at CSP2
Actual Result	PatientX Logged in Successfully at CSP2
Status	Pass
Carried Out On	22-12-2013
Carried Out By	Umme Habiba

Table 2 Test Case 02

Test Case Title	Access Right Delegation Test
Test Case ID	Test 02
Test Objective	To confirm that access rights are delegated from DoctorA to DoctorB.
Pre-Condition	1. DoctorA must log-in. 2. DoctorA must have the access delegation rights on PatientX's medical information. 3. DoctorB has no access rights on PatientX's medical information
Post-Condition	DoctorB should be able to view PatientX's medical information.
Procedure	**DoctorA:** 1. Select **Access Right Delegation** option from the main menu 2. Select *PatientX* as *Subject* 3. Select **Medical Information** as *Resource*. 4. Select **View** as *Action*. 5. Select **24 hours** as *Environment* and click OK. **DoctorB:** 6. Log-in into the system. 7. Click on his list of pateints, PatientX appears to be one of his patients. 8. DoctorB selects PatientX and views his medical information
Expected Result	DoctorB views PatientX's medical information
Actual Result	DoctorB views PatientX's medical information
Status	Pass
Carried Out On	22-12-2013
Carried Out By	Umme Habiba

6 Conclusion

In this work, we have presented the design and architecture of secure IDMS for federated Cloud environment, that helps to eradicate its security and interoperability challenges. Presented system provides typical identity management services along with many advanced features such as *access right delegation, identity federation, real-time synchronization and self-service*. Provisioning, de-provisioning and management of user accounts across multiple Cloud environments is offered through SCIM schema that ensures interoperability while exchanging identity credentials. Implementation of SAML using REST design patterns help ensures the security of CSC's credentials while performing authentication across multiple Cloud environments. In addition to this, proposed system offers a complete implementation of access control system using XACML, that ensures effective authorization and access right delegation. To conclude, presented secure IDMS for federated Cloud environment facilitates the CSPs and CSCs by offering them with secure management and quick access to CSC's credentials that finally improves their business performance. Furthermore, we have configured open-source *Openstack* Cloud infrastructure for the deployment of our system in real-time Cloud scenarios. To evaluate the effectiveness of the implemented functionality, we have followed the guidelines and recommendations given by NIST and CSA and designed various user defined test cases. Results of our analytical evaluation proves that the presented work has ensured the desired features and level of security expected from a secure IDMS for federated Cloud environment.

References

1. Jøsang, A., Fabre, J., Hay, B., Dalziel, J., Pope, S.: Trust requirements in identity management. In: Proceedings of the 2005 Australasian Workshop on Grid Computing and e-research, vol. 44, pp. 99–108. Australian Computer Society, Inc. (2005)
2. Habiba, U., Ghafoor, A., Masood, R., Shibli, M.A.: Assessment criteria for cloud identity management systems. In: 19th IEEE Pacific Rim International Symposium on Dependable Computing (PRDC 2013). IEEE (2014)
3. Fox, A., Griffith, R., Joseph, A., Katz, R., Konwinski, A., Lee, G., Patterson, D., Rabkin, A., Stoica, I.: Above the clouds: A berkeley view of cloud computing. Dept. Electrical Eng. and Comput. Sciences, University of California, Berkeley, Rep. UCB/EECS, vol. 28 (2009)
4. Ghazizadeh, E., Zamani, M., Ab Manan, J.-L., Pashang, A.: A survey on security issues of federated identity in the cloud computing. In: 2012 IEEE 4th International Conference on Cloud Computing Technology and Science (CloudCom), pp. 532–565. IEEE (2012)
5. Angin, P., Bhargava, B., Ranchal, R., Singh, N., Linderman, M., Ben Othmane, L., Lilien, L.: An entity-centric approach for privacy and identity management in cloud computing. In: 2010 29th IEEE Symposium on Reliable Distributed Systems, pp. 177–183. IEEE (2010)
6. Sanchez, R., Almenares, F., Arias, P., Diaz-Sanchez, D., Marín, A.: Enhancing privacy and dynamic federation in idm for consumer cloud computing. IEEE Transactions on Consumer Electronics 58(1), 95–103 (2012)
7. Shamoon, I., Rajpoot, Q., Shibli, A.: Policy conflict management using xacml. In: 2012 8th International Conference on Computing and Networking Technology (ICCNT), pp. 287–291 (August 2012)
8. Celesti, A., Tusa, F., Villari, M., Puliafito, A.: Security and cloud computing: intercloud identity management infrastructure. In: 2010 19th IEEE International Workshop on Enabling Technologies: Infrastructures for Collaborative Enterprises (WETICE), pp. 263–265. IEEE (2010)
9. Yan, L., Rong, C., Zhao, G.: Strengthen cloud computing security with federal identity management using hierarchical identity-based cryptography. In: Jaatun, M.G., Zhao, G., Rong, C. (eds.) Cloud Computing. LNCS, vol. 5931, pp. 167–177. Springer, Heidelberg (2009)
10. Chadwick, D.W., Casenove, M.: Security apis for my private cloud-granting access to anyone, from anywhere at any time. In: 2011 IEEE Third International Conference on Cloud Computing Technology and Science (CloudCom), pp. 792–798. IEEE (2011)
11. Kim, I.K., Pervez, Z., Khattak, A.M., Lee, S.: Chord based identity management for e-healthcare cloud applications. In: 2010 10th IEEE/IPSJ International Symposium on Applications and the Internet (SAINT), pp. 391–394. IEEE (2010)
12. Kumaraswamy, S., Lakshminarayanan, S., Stein, M.R.J., Wilson, Y.: Domain 12: Guidance for identity & access management v2. 1. Cloud Security Alliance 10 (2010), http://www.cloudsecurityalliance.org/guidance/csaguide-dom12-v2
13. Hitachi id identity manager (January 2014), http://hitachi-id.com/identity-manager/ (accessed August 28, 2013)
14. Identity manager 4 standard edition, https://www.netiq.com/products/identity-manager/standard/features/ (accessed August 2013)

15. Morley, M., Lawrence, B.: The cloud: Changing the business ecosystem, http://msdn.microsoft.com/en-us/library/cc836391.aspx (accessed August 28, 2013)
16. Mcafee cloud identity manager, http://www.mcafee.com/ca/resources/data-sheets/ds-cloud-identity-manager.pdf (accessed August 28, 2013)
17. Jansen, W.: Directions in security metrics research. DIANE Publishing (2010)
18. Cremers, C.J.F.: The scyther tool: Verification, falsification, and analysis of security protocols. In: Gupta, A., Malik, S. (eds.) CAV 2008. LNCS, vol. 5123, pp. 414–418. Springer, Heidelberg (2008)

Using Word N-Grams as Features in Arabic Text Classification

Abdulmohsen Al-Thubaity, Muneera Alhoshan, and Itisam Hazzaa

Abstract. The feature type (FT) chosen for extraction from the text and presented to the classification algorithm (CAL) is one of the factors affecting text classification (TC) accuracy. Character N-grams, word roots, word stems, and single words have been used as features for Arabic TC (ATC). A survey of current literature shows that no prior studies have been conducted on the effect of using word N-grams (N consecutive words) on ATC accuracy. Consequently, we have conducted 576 experiments using four FTs (single words, 2-grams, 3-grams, and 4-grams), four feature selection methods (document frequency (DF), chi-squared, information gain, and Galavotti, Sebastiani, Simi) with four thresholds for numbers of features (50, 100, 150, and 200), three data representation schemas (Boolean, term frequency-inversed document frequency, and lookup table convolution), and three CALs (naive Bayes (NB), k-nearest neighbor (KNN), and support vector machine (SVM)). Our results show that the use of single words as a feature provides greater classification accuracy (CA) for ATC compared to N-grams. Moreover, CA decreases by 17% on average when the number of N-grams increases. The data also show that the SVM CAL provides greater CA than NB and KNN; however, the best CA for 2-grams, 3-grams, and 4-grams is achieved when the NB CAL is used with Boolean representation and the number of features is 200.

Keywords: Arabic text classification, feature extraction, classification algorithms, classification accuracy.

Abdulmohsen Al-Thubaity · Muneera Alhoshan
Computer Research Institute, King Abdulaziz City for Science and Technology, Riyadh, KSA
e-mail: {aalthubaity,malhawshan}@kacst.edu.sa

Itisam Hazzaa
College of Computer and Information Sciences, King Saud University, Riyadh, KSA
e-mail: 429202034@student.ksu.edu.sa

© Springer International Publishing Switzerland 2015
R. Lee (ed.), *SNPD*,
Studies in Computational Intelligence 569, DOI: 10.1007/978-3-319-10389-1_3

1   Introduction

1.1   Background

Several initiatives, such as the King Abdullah Initiative for Arabic Content, have been launched in the past few years to support the growth and quality of Arabic content on the Internet. Several studies have advocated the rapid growth of Arabic content on the web as well, as in systems of government institutions and private enterprises [1]. This proliferation of Arabic content requires techniques and tools that are capable of organizing and handling that content in intelligent ways. Text classification (TC)—the process of automatically assigning a text to one or more predefined classes—is one of many techniques that can be used to organize and maximize the benefits of existing Arabic content.

Researchers have focused considerable effort on English TC. Applying existing techniques that were proven to be suitable for English TC may seem to be a simple option for Arabic TC (ATC). However, it has been shown that what is suitable for English TC is not necessarily suitable for Arabic [2].

In general terms, implementing a TC system requires several consecutive steps, collecting a representative text sample, dividing the sample into training and testing sets, extracting the features, selecting the representative features, representing the selected feature for the classification algorithm (CAL), applying the algorithm, producing the classification model, applying the algorithm on testing data, and evaluating the performance of the classification model. The techniques used in each of the above steps affect the accuracy of the TC system in various ways. In this study, we investigate an effect that has not been studied previously in ATC. Specifically, we study the effect of feature type (FT) in ATC accuracy. Instead of single words, word N-grams (N consecutive words) are employed as features using four feature-selection methods, three representation schemas (RSs), and three CALs.

In the rest of this section, we summarize the FTs previously used in ATC and the reported results of using the word N-gram as a feature for TC in other languages. The dataset we have used in this study is illustrated in Section 2.1, and the feature selection (FS) methods, feature RSs, CALs, and other experimental parameters are presented in Section 2.2. In Section 3, the results and our interpretation are discussed. In Section 4, we outline our conclusions and future work.

1.2   Related Work

When planning a TC implementation, the FTs required for text representation must be considered. In ATC, three FTs are primarily used, the first and most common being word orthography (see, for example, [3][4][5]). In this approach, any sequence of Arabic letters bounded by two spaces is considered a feature. In the Arabic writing system, the features produced with this method can be, for example, a single word such as "book" (كتاب), *the* + word "the book" (الكتاب), *two*

+ word "two books" (كتابان/كتابين), or a complete sentence "I will write it" (سأكتبه). Different FS methods can be used to reduce the feature space.

The second type is a stem or root. In this approach, each word in the dataset is analyzed morphologically to remove affixes, extracting the stem of the word orthography, and this stem is then analyzed further to extract its root. This approach is useful for reducing the number of features and the sparseness of data. Usually no FS method is used with this approach. The results show that using word orthography is more accurate for ATC [6][7][8][9]. The reason for the poor results of using stem or root as features for ATC is the low accuracy of the morphology analyzers used [10].

The third type is the character N-gram. In this approach, any consecutive N characters can be considered a feature. This model involves trying to remove affixes without any morphological analysis to get the root/stem, which is three letters for most Arabic words [11][12]. The drawback of this approach is that it produces a very large number of features, possibly affecting classification accuracy (CA). To the best of our knowledge, no comparative analysis has been conducted with the same dataset and experimental environments to assess the performance of using the three feature types.

The combined use of unigram word orthography and bigram word orthography features for ATC was examined in [13]. The authors compared the use of word orthography unigrams and bigrams to the use of word orthography unigrams alone in CA of the k-nearest neighbor (KNN) CAL. They used document frequency (DF) for FS, with a threshold of three, and term frequency-inversed document frequency (TF-IDF) as the RS. They argued that the combined use of word orthography unigrams and bigrams provides greater accuracy than using only single words. We cannot trust this argument fully, because the authors used a subset (four classes) of a dataset of 1,445 texts distributed over nine classes but provided no justification for selecting those four classes, rather than the entire dataset.

Studying the effect of using word-level N-grams on TC for other languages has shown contradictory results. Although use of the single word provides greater accuracy for Turkish TC [14], the data show that using N-grams produces better results than single terms for Farsi TC [15]. To the best of our knowledge, there has been no study that compares the accuracy of ATC using only word-level N-grams as features with that using word orthography. That is what we do in this study.

2   Materials and Methods

2.1   Dataset

We used the Saudi Press Agency (SPA) dataset, a part of the King Abdulaziz City for Science and Technology (KACST) ATC dataset that has been utilized in several ATC studies [9][16] [3][17][18][5]. This dataset consists of 1,526 texts evenly divided into six news classes, cultural, sports, social, economic, political, and general. The basic SPA statistics are illustrated in Table 1.

Table 1 SPA dataset basic statistics

News Classes	No. of Texts	No. of Unique			
		Words	*2-grams*	*3-grams*	*4-grams*
Cultural	258	12,521	33,499	40,311	42,137
Economic	250	9,158	26,305	32,363	34,184
General	255	12,639	32,336	38,384	39,976
Political	250	9,720	26,667	32,136	33,278
Social	258	11,818	33,914	42,279	44,999
Sports	255	8,641	24,425	31,339	33,885
Total	1,526	36,497	154,240	208,390	224,903

2.2 Experimental Setup

As mentioned in Section 1, several consecutive steps must be followed in creating the classification model. For data preprocessing, we removed stop words, which consist primarily of function words of Arabic, such as prepositions and pronouns [5]. In addition, we removed numbers, Latin characters, and diacritics and normalized different forms of Hamza (! ‹|) to (|) and Taa Marbutah (ة) to (ه). We randomly chose 70% of the data for training and the remaining 30% for testing.

We selected single words, 2-grams, 3-grams, and 4-grams as features and chose four of the most used FS methods: DF, chi-squared (CHI), information gain (IG), and Galavotti, Sebastiani, Simi (GSS). The objective of FS is to order the features according to their importance to a given class, with the most appropriate features that distinguish the class from the other classes in the training dataset in the highest ranks. The highest-ranked features are then selected. This selection process yields reduced feature spaces, positively affecting CA and speed. The mathematical representations of DF, CHI, IG, and GSS are shown in Equations (1), (2), (3), and (4), respectively.

$$DF(t, c_i) = D(t, c) \tag{1}$$

$$CHI(t_i, c_j) = \frac{|Tr| \cdot [P(t_i, c_j) \cdot P(\bar{t}_i, \bar{c}_j) - P(t_i, \bar{c}_j) \cdot P(\bar{t}_i, c_j)]^2}{P(t_i) \cdot P(\bar{t}_i) \cdot P(c_j) \cdot P(\bar{c}_j)} \tag{2}$$

$$IG(t, c_i) = \sum_{i=1}^{i=m} P(t, c_i). \log \frac{P(t, c_i)}{P(t).P(c_i)} + \sum_{i=1}^{i=m} P(\bar{t}, c_i) \log \frac{P(\bar{t}, c_i)}{P(\bar{t}).P(c_i)} \tag{3}$$

$$GSS(t, c_i) = P(t, c_i). P(\bar{t}, \bar{c}_i) - P.(t, \bar{c}_i). P(\bar{t}, c_i) \tag{4}$$

where:
m is the number of classes,
Tr is the total number of texts in the training dataset,
c denotes class,
t is a term, which could be a 1-, 2-, 3-, or 4-word level gram,
$D(t, c_i)$ is the number of documents in class c_i that contain term t at least once,

$P(c_i)$ is the probability of class c_i,

$P(t, c_i)$ is the joint probability of class c_i and the occurrence of term t,

$P(t|c_i)$ is the probability of t given c_i.

We used Boolean, TF-IDF, and lookup table convolution (LTC) to choose the way the selected features would be weighted and represented numerically to the CAL. Their mathematical representations are shown in Equations (5), (6), and (7), respectively, where a_{tj} is the numerical weighting of selected feature j in text t:

$$a_{tj} \begin{cases} 1, & \text{if the word exists in the text} \\ 0, & \text{if the word does not exist in the text} \end{cases} \tag{5}$$

$$a_{tj} = f(w) \log\left(\frac{Tr}{d(w)}\right) \tag{6}$$

$$a_{tj} = \frac{\log(f(w)+1)\log\left(\frac{Tr}{d(w)}\right)}{\sqrt{\sum_{i=1}^{i=m}\left[\log(f(w)+1)\log\left(\frac{Tr}{d(w)}\right)\right]^2}} \tag{7}$$

where:

(w) is a word in the text t,

Tr is the total number of texts in the dataset,

m is total number of words in the text,

$f(w)$ is the frequency of the word w_i in the text t and $d(w)$ is the number of texts t that the word w_i occurs in.

An $n \times m$ matrix is then constructed, where n is the number of features, and m is the number of texts in the training dataset. Each cell in this matrix is the weight of feature j in text t. The selected features from the training dataset are then extracted from that dataset and represented in the same way as they were in the training dataset.

Table 2 Experimental parameters

Parameters	Description
Dataset	SPA: 6 classes, 1,526 texts
Training size	70%
Preprocessing	Removing Arabic diacritics, numbers, Latin characters, and stop list words; normalizing Hamza and Taa Marbutah
Testing Size	30%
FTs	Single word, 2-gram, 3-gram, 4-gram
FS methods	DF, CHI, IG, GSS
No. of Terms Selected	High-ranked terms (50, 100, 150, 200)
Threshold	Minimum DF = 10
FR schemas	Boolean, TF-IDF, LTC
CALs	NB, KNN, SVM
Number of experiments	576

To train the CAL using the selected features from the training data, we used three of the most commonly used CALs, KNN, naive Bayes (NB), and support vector machine (SVM). For more information regarding FS methods, feature RSs, and CALs, see [19]. We used the RapidMiner 4.0 [20] implementations of these CALs to train and test the classification model. The experimental parameters are summarized in Table 2.

3   Results and Discussion

Conducting experiments with the parameters in Table 2, we obtained the data in Table 3.

Table 3 Experimental results

Classifier	Grams	Average	Minimum		Maximum	
NB	1	60.35	39.25	IG, LTC, 150	75	CHI, TFiDF, 200
	2	50.38	25.88	CHI, LTC, 200	66.45	IG, Boolean, 200
	3	38.76	21.05	CHI, LTC, 200	51.32	DF, Boolean, 200
	4	32.77	18.86	CHI, LTC, 200	42.98	GSS, Boolean, 200
KNN	1	49.14	40.13	DF, TFiDF, 100	58.77	CHI, LTC, 50
	2	41.35	28.29	IG, Boolean, 50	51.54	CHI, LTC, 50
	3	37.63	33.33	DF, TFiDF, 50	41.89	IG, LTC, 150
	4	33.54	30.48	IG, Boolean, 50	36.4	GSS, LTC, 150
SVM	1	72.35	67.54	DF, TFiDF, 50	75.44	IG, LTC, 200
	2	58.73	49.78	DF, Boolean, 50	65.13	IG, LTC, 200
	3	41.81	35.53	CHI, Boolean, 50	47.37	IG, TFiDF, 200
	4	35.07	31.58	CHI,Boolean,50	38.6	GSS, LTC, 100

The table lists the average CA for each gram number of each classifier, the minimum and maximum CA for that classifier, and the combination of FS method, RS, and number of terms that has produced the respective value. The data suggest that, on average, for all gram numbers, the CA of SVM is greater than that of NB, followed by that of KNN. In addition, the data suggest that greater CA is achieved when single words are used as a feature, and that CA declines by 17%, on average, when the gram number increases.

Notably, while SVM achieved greater CA, on average, and the best CA using single words, NB exhibited greater CA for 2-grams, 3-grams, and 4-grams. The data show that the best CA was achieved when the number of terms was 200, the maximum number of terms used in our experiments. According to the data, NB worked well with Boolean representation (three of the best results were achieved with Boolean) and not well with LTC (all the worst classification results were achieved with LTC); the case was the opposite for SVM. The best results for KNN were achieved using LTC with 50 features. In all the experiments, CA using single words as a feature was greater than when using 2-grams, 3-grams, and 4-grams, except for one case. In that case, in which KNN, CHI, and TF-IDF were used with 150 terms, single words achieved 48.46 while 2-grams achieved 49.78.

Table 4 outlines additional results from using the best combination of parameters yielding the greatest CA (75.44). The table shows the CA of SVM using IG for FS and LTC as an RS for all gram numbers and all numbers of features.

Table 4 Classification accuracy using SVM, IG, and LTC

Number of Features	1-gram	2-grams	3-grams	4-grams
50	71.05	58.55	38.38	33.55
100	74.78	62.72	39.91	34.21
150	75.00	62.72	42.76	37.5
200	75.44	65.13	45.39	35.53

The data illustrate two main observations. The first is that CA increases when the number of features increases. The second is that CA decreases when the gram number increases. Accuracy falls by approximately 16%, 33%, and 15%, on average, when 2-grams are used compared to single words, when 3-grams are used compared to 2-grams, and when 4-grams are used compared to 3-grams, respectively. Figure 1 depicts the decrease in CA, when the gram number increases.

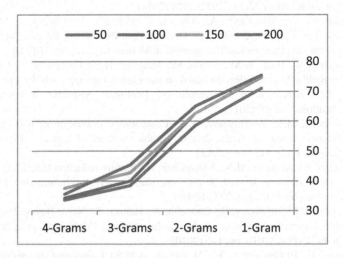

Fig. 1 Increase in CA with decrease in gram number

One of the reasons for the poor CA results, when N-grams were used as features, was that the data were sparse, as a result the facts that the same Arabic word can occur in different forms and that using word N-grams increases the number of features, decreasing the frequency of each feature.

4   Conclusion

In this study, we conducted experiments to investigate the effect of word N-grams as features for ATC. We used the SPA dataset, four FS methods, different numbers of top-ranked features, three feature RSs, and 3 CALs, as summarized in Table 2. The data, summarized in Table 3, show that single words as a feature for ATC produced better results than other word N-grams. The best average CA was achieved using the SVM CAL; however, the NB CAL provided better CA results for 2-grams, 3-grams, and 4-grams compared to SVM and KNN, using the same word N-grams.

The main conclusion of our study is that the use of single words for ATC is more effective than using N-grams. Further investigation is required in our future work to validate our results using larger datasets from different domains and genres, such as newspapers and scientific texts, to reduce data sparseness.

References

1. Alarifi, A., Alghamdi, M., Zarour, M., Aloqail, B., Lraqibah, H., Alsadhan, K., Alkwai, L.: Estimating the Size of Arabic Indexed Web Content. Scientific Research and Essays 7(28), 2472–2483 (2012)
2. Mesleh, A.M.: Feature sub-set selection metrics for Arabic text classification. Pattern Recognition Letters 32(14), 1922–1929 (2011)
3. Althubaity, A., Almuhareb, A., Alharbi, S., Al-Rajeh, A., Khorsheed, M.: KACST Arabic Text Classification Project: Overview and Preliminary Results. In: 9th IBMIA Conference on Information Management in Modern Organizations (2008)
4. Alwedyan, J., Hadi, W.M., Salam, M., Mansour, H.Y.: Categorize Arabic data sets using multi-class classification based on association rule approach. In: Proceedings of the 2011 International Conference on Intelligent Semantic Web-Services and Applications, vol. 18 (2011)
5. Khorsheed, M.S., Al-Thubaity, A.O.: Comparative evaluation of text classification techniques using a large diverse Arabic dataset. Language Resources and Evaluation 47(2), 513–538 (2013)
6. Duwairi, R., Al-Refai, M.N., Khasawneh, N.: Feature reduction techniques for Arabic text categorization. Journal of the American Society for Information Science and Technology 60(11), 2347–2352 (2009)
7. Noaman, H.M., Elmougy, S., Ghoneim, A., Hamza, T.: Naive Bayes classifier based Arabic document categorization. In: 7th International Conference on Informatics and Systems (INFOS 2010), pp. 1–5 (2010)
8. Harrag, F., El-Qawasmah, E., Al-Salman, A.M.S.: Comparing dimension reduction techniques for Arabic text classification using BPNN algorithm. In: First International Conference on Integrated Intelligent Computing (ICIIC 2010), pp. 6–11 (2010)

9. Al-Shammari, E.T.: Improving Arabic document categorization: Introducing local stem. In: 10th International Conference on Intelligent Systems Design and Applications (ISDA), pp. 385–390 (2010)
10. Sawalha, M., Atwell, E.S.: Comparative evaluation of Arabic language morphological analysers and stemmers. In: Proceedings of COLING 2008 22nd International Conference on Computational Linguistics (Poster Volume), pp. 107–110. Coling 2008 Organizing Committee (2008)
11. Sawaf, H., Zaplo, J., Ney, H.: Statistical classification methods for Arabic news articles. In: Proceedings of the ACL/EACL 2001 Workshop on Arabic Language Processing: Status and Prospects, Toulouse, France (2001)
12. Khreisat, L.: A machine learning approach for Arabic text classification using N-gram frequency statistics. Journal of Informetrics 3(1), 72–77 (2009)
13. Al-Shalabi, R., Obeidat, R.: Improving KNN Arabic text classification with n-grams based document indexing. In: Proceedings of the Sixth International Conference on Informatics and Systems, Cairo, Egypt, pp. 108–112 (2008)
14. Güran, A., Akyokucs, S., Bayazit, N.G., Gürbüz, M.Z.: Turkish text categorization using N-gram words. In: Proceedings of the International Symposium on Innovations in Intelligent Systems and Applications (INISTA 2009), pp. 369–373 (2009)
15. Bina, B., Ahmadi, M., Rahgozar, M.: Farsi text classification using n-grams and KNN algorithm: A comparative study. In: Proceedings of the 4th International Conference on Data Mining (DMIN 2008), pp. 385–390 (2008)
16. Froud, H., Lachkar, A., Ouatik, S.A.: A comparative study of root-based and stem-based approaches for measuring the similarity between Arabic words for Arabic text mining applications. arXiv preprint arXiv:1212.3634 (2012)
17. Al-Harbi, S., Almuhareb, A., Al-Thubaity, A., Khorsheed, M., Al-Rajeh, A.: Automatic Arabic text classification. In: 9es Journées Internationales d'Analyse Statistique des Données Textuelles, JADT 2008, pp. 77–83 (2008)
18. Al-Saleem, S.: Associative classification to categorize Arabic data sets. International Journal of ACM JORDAN 1, 118–127 (2010)
19. Sebastiani, F.: Machine learning in automated text categorization. ACM Computing Surveys 34(1), 1–47 (2002)
20. Mierswa, I., Wurst, M., Klinkenberg, R., Scholz, M., Euler, T.: YALE: Rapid prototyping for complex data mining tasks. In: Ungar, L., Craven, M., Gunopulos, D., Eliassi-Rad, T. (eds.) KDD 2006 Proceedings of the 12th ACM SIGKDD International Conference on Knowledge Discovery and Data Mining, pp. 935–940. ACM, New York (2006)

Optimization of Emotional Learning Approach to Control Systems with Unstable Equilibrium

Mohammad Hadi Valipour, Khashayar Niki Maleki, and Saeed Shiry Ghidary

Abstract. The main problem concerning model free learning controllers in particular BELBIC (Brain Emotional Learning Based Intelligent controller), is attributed to initial steps of learning process since the system performance is dramatically low, because they produce inappropriate control commands. In this paper a new approach is proposed in order to control unstable systems or systems with unstable equilibrium. This method is combination of one imitation phase to imitate a basic solution through a basic controller and two optimization phases based on PSO (Particle Swarm Optimization) which are employed to find a new solution for stress generation and to improve control signal gradually in reducing error. An inverted pendulum system is opted as the test bed for evaluation. Evaluation measures in simulation results show the improvement of error reduction and more robustness than a basic tuned double-PID controller for this task.

1 Introduction

Traditional control methods are based on system identification, modeling, and designing controller regarding to predefined goals for under control systems. As these systems become more complex, on one hand their identification grow to be much more difficult and sometimes impossible, and on the other hand there are some factors such as dynamics of system, uncertainty, and decay induced system changes, that require redesigning or readjusting the controller [15]. Readjusting parameters

Mohammad Hadi Valipour · Saeed Shiry Ghidary
Department of Computer Engineering, Amirkabir University of Technology,
Tehran 158754413, Iran
e-mail: {valipour,shiry}@aut.ac.ir

Khashayar Niki Maleki
The University of Tulsa, Tulsa, OK 74104, United States
e-mail: niki-maleki@utulsa.edu

© Springer International Publishing Switzerland 2015   45
R. Lee (ed.), *SNPD*,
Studies in Computational Intelligence 569, DOI: 10.1007/978-3-319-10389-1_4

of controller itself is a quite time consuming process, even for model free ones like fuzzy controllers we need to adjust the parameters automatically or manually [4]. Therefore, adaptive and intelligent control are new approaches in control engineering. In intelligent control there is an immense desire to inspire from natural systems, by way of illustration, we can refer to Neuro-Fuzzy [7], evolutionary systems [22], intelligent controllers based on Reinforcement Learning [8], and Multi Agent controllers [23]. Advanced findings in Neuroscience indicate that emotion plays a significant role in human reasoning and decision making [18]. Recently, emotion has also introduced to artificial intelligence as a significant factor in decision making of expert systems [18]. Humans and other mammals are known as creatures which have emotional behaviors. The emotional decision making process which usually takes place fast, helps mammals keep safe in dangerous situations. LIMBIC is a part of brain in mammals which controls emotional process. This system is been modeled mathematically in [1][18]. AMYGDALA and ORBITOFRONTAL are two main parts of limbic, and their models were initially introduced in [19]. Brain Emotional Learning Based Intelligent Controller (BELBIC) introduced as an intelligent controller [15] in result of attempts and previous works in [19]. BELBIC intelligent controller is developed based on middle brain of mammals called BEL [19]. This controller has been successfully applied to many simulated control problems [14][17]; moreover, it has been utilized in many real problems, and have had acceptable results in comparison other control methods[10]. Additionally we can appoint many other successful applications of this controller such as: HVAC Systems [26], Robocup issues [24][25], Control of Intelligent Washing Machines [16], Speed Control of Switched Reluctance Motor [20], Active Queue Management [9], Model Free Control of Overhead Travelling Crane [10]. The main problem concerning learning controllers which do not need any background knowledge of system dynamics nor system models in particular RL based controllers and BELBIC, is that in initial steps of learning process because of producing false control commands, they might cause low performance. If this part of learning does not lead to in instability, the controller can learn the proper control signals progressively otherwise or in the case that the system is innately unstable, implementing these controllers may cause the system become unstable and the process must be stopped. Even though BELBIC demonstrates fast learning ability, it shares the same problem and cannot be applied to such systems [11]. To overcome similar difficulties another approach is introduced. In this paper an evolutionary method based on Particle Swarm Optimization (PSO) [13] is introduced to improve control signal and reduce error gradually. The sensory signal generation unit is a combination of error and its derivation with constant coefficients. Producing proper stress signal is an important factor in this section. In brief, in this approach initially, BELBIC controller output must become similar to a preliminary controller. At this level, the stress signal generation unit can be a weighted sum of some characteristics of error. After early learning, generation of initial signal will turn to a new weighted sum with 6 weights. The next step is enhancing stress signal in order to reduce BELBIC controller error in the time of controlling inverse pendulum. This enhancement will be performed by adjusting the weights of stress generation neural network through PSO. This approach has demonstrated outstanding performance which is presented in simulation results. This paper is organized as follows. Section 2 reviews intelligent control based on brain emotional learning. Section 3 introduces its computation model. In Section 4 the structure of our implemented controller is discussed. Section 5 is about stress generation. Simulation and experimental results are given in Section 6. Finally, conclusion and future works are presented in section 7.

2 Control Based on Brain Emotional Learning

Decision making process in human brains is not limited to computation and logic which take place in upper portion of brain, also emotions that their source is in cerebellum and middle portion of brain are involved in decision making. Indeed solving a decision making problem considering the complexity of the solution based on computation and logic besides the existence of uncertainty sometimes is too intricate. Accordingly, before the problem is processed in cognitive stage, it means considering total representation of external stimulus, the process of problem would be performed with simpler representation of stimulus by emotions and relatively good answer would be obtained so fast [4]. That is how emotional process accelerates decision making procedure. There were many attempts for identification of emotional decision making process In 80s which led to emotional system to be introduced as an expert system [5]. In this approach images that involve the representation of stimulus and the response of the expert system to it will be labeled with good or bad and during decision making those images which are labeled bad will be omitted and decision making will takes place between the rest of images [5][27]. In recent approaches, presenting the computational model of those brain segments which are responsible for emotional process is been considered. In methods based on computational models, emotions are signals representing outer environment. In psychological researches emotion is introduced as desirability degree factor [6]. There are the same approaches in Control. In proposed controller in [12] those items that designer is sensitive to them are considered as stimulations which cause stress, and control system must work in the way that reduces the stress. Therefore, an agent based neuro-fuzzy controller is designed in [7] in which the parameters are adjusted by learning. Moreover, the fuzzy controller proposed in [21] employs 3 negative emotions: fear, anxiety, and pain for learning a mobile robot. In Brain Emotional Learning Based Intelligent Controller (BELBIC) [15] which is remarked in this paper, stress would be produced as a negative factor and the parameters of controller which has a network structure would be adjusted based upon it. This controller is founded based on AMYGDALA computational model [19].

3 Computation Model of Brain Emotional Learning

Amygdala is a part of brain which is responsible for emotional processes and is connected to sensory layer, Thalamus, and Orbitofrontal portion (Fig. 1). Amygdala and Orbitofrontal have network architecture in their computational models in which

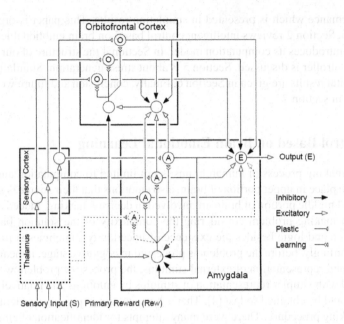

Fig. 1 Amygdala model of Brain Emotional Learning (BELBIC structure)

there is a node for each sensory input. There is another node in Amygdala for Thalamus input, the amount of it is equal to maximum of sensory inputs. Output of nodes in Amygdala and Orbitofrontal is calculated according to equations (1) and (2).

$$A_i = S_i \times V_i \qquad (1)$$

$$O_i = S_i \times W_i \qquad (2)$$

Where A_i and O_i are outputs of nodes in network structure of Amygdala and Orbitofrontal respectively, V and W are weights of nodes and S_i is sensory input. Changes in V and W in learning process are calculated by equations (3) and (4) respectively.

$$\Delta V_i = \alpha \times max(0, S_i \times (R - \sum_j A_j)) \qquad (3)$$

$$\Delta W_i = \beta \times S_i \times (\sum_j O_j - R) \qquad (4)$$

As it is shown, amount of A_i cannot be reduced which means that forgetting learned information wont take place in Amygdala. In fact forgetting or lets say prevention is Orbitofrontal duty. Finally model output obtains from equation (5).

$$E = \sum_i A_i - \sum_i O_i \qquad (5)$$

There is one more input to Amygdala, A_{th}, which is the maximum values of sensory inputs:

$$A_{th} = max(S_i) \qquad (6)$$

Figure 2 presents BELBIC controller. Ultimately α and β, initial values of A, A_{th}, O, and functions S (Sensory Generator) and R (Stress Generator) in producing emotional signal must be selected properly.

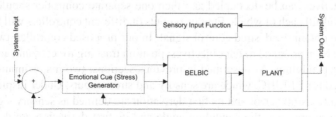

Fig. 2 Control system configuration using BELBIC

4 Design and Structure of the Controller

Control task of unstable systems or stable systems with unstable equilibrium is one of most difficult and interesting problems in control issues. Usually, controlling inverted pendulum, as one of the problems which is classified in this category, is utilized for testing and reviewing of proposed approaches in such systems. This problem is described as tracking reference signal and stabilizing the pendulum at the same time. Due to nonlinearity of systems state equations, it is not easy to design a model based controller for this system. In this paper BELBIC is employed as a model free controller. One of the BELBIC properties is fast learning, however there is no information about the dynamic of system and the pendulum may falls down in primary steps; consequently the learning process terminates soon. It means that BELBIC should learn a proper control signal in the short time regarding to sensory and stress input signals, but due to sensitivity of pendulum angle to error, even a little error leads system to be unstable. In simulation this leaning process takes too long and probably impossible in real applications [11].

We have proposed a multi phase solution to solve this problem. In phase one, an imitative process takes place. BELBIC controller learns the control signal produced by another basic controller (a proper tuned double PID for this task) which can stabilize the inverted pendulum. In the next phase, we explore to find the first solution better than above basic controller ,and finally in the last phase, the solution of phase two will be improved. The first phase will be finished in short time and networks' weights will be updated and fixed. The second phase is designed to find the first better solution and to replace it with current basic controller, so we the stress generation method will be changed to the new structure and explores the solution space. This method employes particle swarm optimization algorithm in order to find one stress generation block that leads to more error reduction. In this phase stress generation block changes to a weighted sum of six components, then the candidate solution will be considered as the best solution in solution space. In the third phase the optimization process will be continued to reduce error.

4.1 Controller Structure Details

To design a controller which satisfies different objectives at first we assume that these objectives can be decoupled and then one separate controller should be designed to fullfill each of which. Finally, outputs of different controllers will be fused together to obtain the desired control signal. In our proposed controller, two BEL-BICs have been assigned to our objectives: position tracking for cart, and angle regulating for pendulum. As described in previous sections, two major inputs should be provided for BELBIC, which are sensory and stress inputs. Under inspiration of [11] the cart position error and its first derivation are defined as sensory signals for one Belbic; more over, the pendulum angle and its first derivation are defined as sensory signals for the other one. Note that each BELBIC controller has two neurons for sensory inputs. Fig. 3 shows the design and structure of proposed controller in Simulink environment. As indicated in this figure, just a summation operator is used to combine the output of controllers. Decoupling of controllers is compensated by fusing the objectives in reward or stress functions. This approach has two major benefits: first, reduction of fusion cost, and second, not changinge the structure of controller between learning and optimization phases.

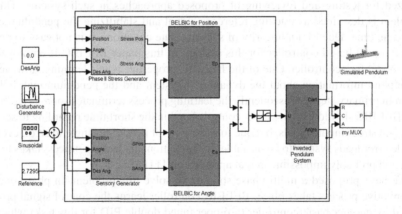

Fig. 3 Design and structure of proposed controller in Simulink environment

5 Stress Generation

5.1 Imitation Phase

In imitation phase, BELBIC learns the control signal produced by the initial basic controller; therefore we need to consider the control signal error to generate stress signal. In this work, a weighted sum of two components are utilized: control signal error and its first derivation. Equation 7 shows the proposed formula to generate stress in this phase.

$$stress_{imit} = w_1 e_u + w_2 \dot{e}_u \tag{7}$$

5.2 Optimization Phase

In optimization phases, we use particle swarm optimization to improve stress generation after imitation phase, in this step the stress signal is obtained by equation 8, and the goal is to find proper weights (w_1 to w_6).

$$stress_{opt} = w_1 e_a + w_2 \dot{e}_a + w_3 e_p + w_4 \dot{e}_p + w_5 s_{cf} + w_6 \dot{s}_{cf} \tag{8}$$

in above equation e_a and $\dot{e}_a$ are angle error and its first derivation, e_p and $\dot{e}_p$ are position error and its first derivation and s_{cf} and $\dot{s}_{cf}$ are control force signal and its first derivation.

In previous sections we mentioned that there are 2 phase for optimization. In first phase of optimization we run 10 BELBICs which have been learned by a basic controller in parallel as our particles and one basic controller as the best candidate for proper controller. After each iteration of optimization (including one sinusoidal period with disturbance), controller will be evaluated by their error reduction and if there is a better solution than basic controller, we will consider it as the best candidate; else the next iteration will be done. After some iteration if we find a new candidate as the best solution we will switch to second phase of optimization. In this phase only 10 BELBICs run in parallel as desired particles and the stress generated by the best particle will be improved gradually.

5.2.1 Particle Swarm Optimization

Particle Swarm Optimization (PSO) is a technique used to explore the search space of a given problem to find the settings or parameters required to maximize a particular objective. This technique, first described by James Kennedy and Russell C. Eberhart in 1995 [13], originates from two separate concepts: the idea of swarm intelligence based on the observation of swarming habits by certain kinds of animals (such as birds and fishes) and the field of evolutionary computation. The PSO algorithm consists of just three steps, which are repeated until some stopping condition is met [2]: First, evaluation of the fitness of each particle, second, updating individual and global best fitnesses and positions and third, update velocity and position of each particle. [3]

The first two steps are fairly trivial. Fitness evaluation is conducted by supplying the candidate solution to the objective function. Individual and global best fitnesses and positions are updated by comparing the newly evaluated fitnesses against the previous individual and global best fitnesses, and replacing the best fitnesses and positions as necessary. The velocity and position update step is responsible for the

optimization ability of the PSO algorithm. The velocity of each particle in the swarm
is updated using the following equation [3]:

$$v_i(t+1) = wv_i(t) + c_1r_1[\hat{x}_i(t) - x_i(t)] + c_2r_2[g_i(t) - x_i(t)] \tag{9}$$

The index of the particle is represented by i. Thus, $v_i(t)$ is the velocity of particle i
at time t and $x_i(t)$ is the position of particle i at time t. The parameters w, c_1, and c_2
are user-supplied coefficients. The values r_1 and r_2 are random values regenerated
for each velocity update. The value $\hat{x}_i(t)$ is the individual best candidate solution
for particle i at time t, and $g(t)$ is the swarms global best candidate solution at time
t. The term $c_1r_1[\hat{x}_i(t) - x_i(t)]$, called the *cognitive component*, acts as the particles
memory, causing it to tend to return to the regions of the search space in which it has
experienced high individual fitness and the term $c_2r_2[g_i(t) - x_i(t)]$, called the *social
component*, causes the particle to move to the best region the swarm has found so far.
Once the velocity for each particle is calculated, each particles position is updated
by applying the new velocity to the particles previous position [3]:

$$x_i(t+1) = x_i(t) + v_i(t+1) \tag{10}$$

This process is repeated until some stopping condition is satisfied.

5.2.2 Particle Structure

As mentioned before, the goal is to find proper weights for stress generation, so
proposed particle structure is a vector of six components including w_1 to w_6 form
equation 8.

6 Simulation and Results

The proposed controller has been compared with the original basic controller, which
is a double PID. Note that without employing imitation phase, simulated BELBIC
controller did not learn the proper control signal in a reasonable time and the pendulum fell down. Also a random voltage produced by a Gaussain distribution has
been applied to the test to evaluate the controller robustness in the presence of disturbance. At the first step of disturbance evaluation, we applied the declared disturbance in 5 times after imitation phase, and in the second step the disturbance
produced by new settings has been applied to the desired cart position whole training duration. Note that in optimization phase the parameters w, c_1, and c_2 are set as
0.5, 1.5 and 1.5 respectively.

Fig. 4 and Fig. 5 show the result of controller in above evaluation tests respectively. In both of them the imitation phase is in first 28.4 seconds. As it can be seen
the BELBIC can imitate the behavior of basic controller in a reasonable time and
the optimization phase is started in this time.

As mentioned before, in the first test we apply the disturbance with zero mean
and 0.2 of variance in some randomly selected times. In this step, the better solution

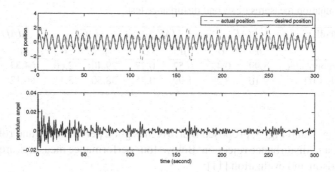

Fig. 4 Results of proposed controller in first disturbance test

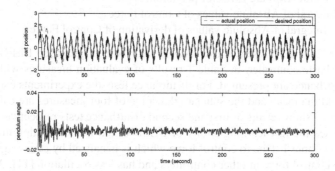

Fig. 5 Results of proposed controller in second disturbance test

Table 1 Evaluation measures without disturbance

Controller	IAE_p	IAE_a	IACF	IADCF
Proposed controller	2.16	0.28	7.74	6.91
Double PID	2.73	0.39	11.02	11.81

Table 2 Evaluation measures of first disturbance test

Controller	IAE_p		IAE_a		IACF		IADCF	
	μ	σ	μ	σ	μ	σ	μ	σ
Proposed controller	5.73	0.71	0.38	0.11	55.18	2.02	94.08	5.96
Double PID	14.91	3.26	1.96	0.40	73.60	2.96	126.03	5.83

than basic controller is found in time 100.2 and the optimization is continued to 217.4. In the second disturbance test a random voltage produced by a Gaussain distribution with zero mean and 0.05 of variance is applied to desired cart position in optimization phase. The better candidate solution for stress generation is found in 100.2 as same as previous test, and the optimization is continued to 217.4 too. As it

Table 3 Evaluation measures of second disturbance test

Controller	IAE_p		IAE_a		$IACF$		$IADCF$	
	μ	σ	μ	σ	μ	σ	μ	σ
Proposed controller	3.99	0.45	0.52	0.27	29.40	2.60	64.33	3.10
Double PID	9.10	3.28	1.99	0.35	38.50	3.11	96.12	3.15

shown in Fig. 4 and Fig. 5, proposed BELBIC has good performance and robustness in tracking and disturbance rejection. Below, four performance measures are defined for comparison and evaluation [11]:

- IAE_a: Integral Absolute Error for cart position
- IAE_p: Integral Absolute Error for pendulum angle
- $IACF$: Integral of Absolute values of Control Force
- $IADCF$: Integral of Absolute values of derivation of Control Force

Tables 1, 2, and 3 present above measures for both of proposed BELBIC controller and basic controller. In Table 1 results of evaluation measures without applying disturbance are presented. For disturbance tests the experiments carried out 10 times and the mean and the standard deviation of four measures are calculated. Tables 2 and 3 show results of first and second disturbance tests respectively.

We can see the fast learning ability of BELBIC and less oscillation than basic controller in Table 1. Due to control force which is penalized by stress signal, it is lower than control force in other controllers and has less oscillation [11]. As it can be seen in Tables 2 and 3, in presence of disturbance, proposed BELBIC is more robust and shows better disturbance rejection.

7 Conclusion and Future Works

When a learning controller such as BELBIC is used to control unstable systems or system with unstable equilibrium, without any background knowledge of system dynamics, they might cause low performance in initial steps of learning process because of producing false control commands. In this paper we proposed a novel approach in optimization of brain emotional learning based intelligent controller to control systems with unstable equilibriums. The proposed approach takes place in three phases, in the first phase BELBIC learns to produce a proper control signal in an imitation process from a basic controller. In the next two phases, an optimization improves the stress which is produced by emotional cue generator. In the second phase, particle swarm optimization finds a new solution having much better performance rather than basic controller considering position and angle error, and eventually, in the third phase, the optimization process continues between candidate solutions to reduce errors. To have an efficient performance evaluation three experiments have been performed. Proposed controller shows better performance than basic controller in all experiments, especially in disturbance tests. Because of learning capability, good disturbance rejection and robustness of BELBIC, it can be employed in many tasks which have unstable equilibriums as a model free controller, such as humanoid robot stability control, self-balancing electric vehicles or two wheel medical transporters.

References

1. Balkenius, C., Moren, J.: A computational model of emotional conditioning in the brain. In: Proceedings of the Workshop on Grounding Emotions in Adaptive Systems, Zurich (1998)
2. van den Bergh, F.: An analysis of particle swarm optimizers. Ph.D. thesis, University of Pretoria (2001)
3. Blondin, J.: Particle swarm optimization: A tutorial (2009)
4. Custodio, L., Ventura, R., Pinto-Ferreira, C.: Artificial emotions and emotion-based control systems. In: 1999 7th IEEE International Conference on Emerging Technologies and Factory Automation Proceedings (ETFA 1999), Barcelona, Spain (1999)
5. Damasio, A.R.: Descartes Error: Emotion, Reason and the Human Brain. G.P. Putnams Sons, New York (1994)
6. El-Nasr, M., Yen, J.: Agents, emotional intelligence and fuzzy logic. In: Proc. 1998 Conference of the North American Fuzzy Information Processing Society (NAFIPS), Pensacola Beach, FL, USA, pp. 301–305 (1998)
7. Fatoorechi, M.: Deveolpement of emotional learning methods for multi modal and multi variable problem. Master's thesis, ECE Department, University of Tehran
8. Hwang, K.S., Tan, S., Tsai, M.C.: Reinforcement learning to adaptive control of nonlinear systems. IEEE Transactions on Systems, Man, and Cybernetics, Part B: Cybernetics 33(3), 514–521 (2004)
9. Jalili-Kharaajoo, M.: Application of brain emotional learning based intelligent controller (belbic) to active queue management. In: Bubak, M., van Albada, G.D., Sloot, P.M.A., Dongarra, J. (eds.) ICCS 2004. LNCS, vol. 3037, pp. 662–665. Springer, Heidelberg (2004)
10. Jamali, M.R., Arami, A., Hosseini, B., Moshiri, B., Lucas, C.: Real time emotional control for anti-swing and positioning control of simo overhead traveling crane. International Journal of Innovative Computing, Information and Control 4(9), 2333–2344 (2008)
11. Javan-Roshtkhari, M., Arami, A., Lucas, C.: Emotional control of inverted pendulum system: A soft switching from imitative to emotional learning. In: Proceedings of the 4th International Conference on Autonomous Robots and Agents, Wellington, New Zealand, pp. 651–656 (2009)
12. Jazbi, A.: Development of reinforcement learning in intelligent controllers and its applications in industries and laboratories. Master's thesis, ECE Department, University of Tehran (1998)
13. Kennedy, J., Eberhart, R.: Particle swarm optimization. In: Proceedings of IEEE International Conference on Neural Networks, pp. 1942–1948 (1995)
14. Lucas, C., Milasi, R.M., Araabi, B.N.: Intelligent modeling and control of washing machine using llnf modeling and modified belbic. Asian Journal of Control 8(4), 393–400 (2005)
15. Lucas, C., Shahmirzadi, D., Sheikholeslami, N.: Introducing belbic: Brain emotional learning based intelligent controller. Intelligent Automation and Soft Computing 10(1), 11–22 (2004)

16. Milasi, R.M., Lucas, C., Araabi, B.N.: Intelligent modeling and control of washing machines using llnf modeling and modified belbic. In: Proc. of International Conference on Control and Automation (ICCA 2005), Budapest, vol. 2, pp. 812–817 (2005)
17. Milasi, R.M., Lucas, C., Arrabi, B.N., Radwan, T.S., Rahman, M.A.: Implementation of emotional controller for interior permanent magnet synchronous motor drive. In: Proc. of IEEE / IAS 41st Annual Meeting: Industry Applications, Tampa, Florida, USA (2006)
18. Moren, J.: Emotion and learning: A computational model of the amygdale. Ph.D. thesis, Lund university, Lund, Sweden (2002)
19. Moren, J., Balkenius, C.: A computational model of emotional learning in the amygdala: From animals to animals. In: Proc. of 6th International Conference on the Simulation of Adaptive Behavior, pp. 383–391. MIT Press, Cambridge (2000)
20. Rashidi, F., Rashidi, M., Hashemi-Hosseini, A.: Speed regulation of dc motors using intelligent controllers. In: Proceedings of 2003 IEEE Conference on Control Applications (CCA 2003), vol. 2, pp. 925–930 (2003)
21. Seif El-Nasr, M., Skubic, M.: A fuzzy emotional agent for decision-making in a mobile robot. In: The 1998 IEEE International Conference on Fuzzy Systems Proceedings, IEEE World Congress on Computational Intelligence, Anchorage, AK, USA, pp. 135–140 (1998)
22. Shahidi, N., Esmaeilzadeh, H., Abdollahi, M., Lucas, C.: Memetic algorithm based path planning for a mobile robot. International Journal of Information Technology 1(2) (2004)
23. Shahidi, N., Gheiratmand, M., Lucas, C., Esmadizade, H.: Utmac: A c ++ library for multi-agent controller design. In: World Automation Congress Proceedings, pp. 287–292 (2004)
24. Sharbafi, M.A., Lucas, C.: Designing a football team of robots from beginning to end. International Journal of Information Technology 3(2), 101–108 (2006)
25. Sharbafi, M.A., Lucas, C., Toroghi Haghighat, A., Amirghiasvand, O., Aghazade, O.: Using emotional learning in rescue simulation environment. Transactions of Engineering, Computing and Technology (2006)
26. Sheikholeslami, N., Shahmirzadi, D., Semsar, E., Lucas, C., Yazdanpanah, M.J.: Applying brain emotional learning algorithm for multivariable control of hvac systems. J. Intell. Fuzzy Syst. 17(1), 35–46 (2006)
27. Ventura, R.M.M., Pinto-Ferreira, C.A.: Emotion-based control systems. In: Proceedings of the 1999 IEEE International Symposium on Intelligent Control/Intelligent Systems and Semiotics, Cambridge, MA, USA, pp. 64–66 (1999)

An Evaluation of the MiDCoP Method for Imputing Allele Frequency in Genome Wide Association Studies

Yadu Gautam, Carl Lee, Chin-I Cheng, and Carl Langefeld

Abstract. A genome wide association studies require genotyping DNA sequence of a large sample of individuals with and without the specific disease of interest. The current technologies of genotyping individual DNA sequence only genotype a limited DNA sequence of each individual in the study. As a result, a large fraction of Single Nucleotide Polymorphisms (SNPs) are not genotyped. Existing imputation methods are based on individual level data, which are often time consuming and costly. A new method, the Minimum Deviation of Conditional Probability (MiDCoP), was recently developed that aims at imputing the allele frequencies of the missing SNPs using the allele frequencies of neighboring SNPs without using the individual level SNP information. This article studies the performance of the MiDCoP approach using association analysis based on the imputed allele frequency by analyzing the GAIN Schizophrenia data. The results indicate that the choice of reference sets has strong impact on the performance. The imputation accuracy improves if the case and control data sets are imputed using a separate but better matched reference set, respectively.

Keywords: Association Tests, Conditional Probability, Imputation, Minimum Deviation, Multilocus Information Measure, Single Nucleotide Polymorphisms.

Yadu Gautam · Carl Lee · Chin-I Cheng
Department of Mathematics, Central Michigan University, Mt. Pleasant, MI 48859, USA
e-mail: {gautalyn,lee1c,cheng3c}@cmich.edu

Carl Langefeld
Department of Biostatistical Sciences, Division of Public Health Sciences,
Wake Forest University, USA
e-mail: clangefe@wfubmc.edu

© Springer International Publishing Switzerland 2015   57
R. Lee (ed.), *SNPD*,
Studies in Computational Intelligence 569, DOI: 10.1007/978-3-319-10389-1_5

1   Introduction

A genome wide association study aims at investigating the association between genetic variations and particular diseases by analyzing the association based on Single Nucleotide Polymorphisms (SNPs) between individuals with the specific disease of interest (case group) and individuals without the disease (control group). The association statistics and the corresponding p-values between case and control based on SNPs are used to determine if these SNPs are statistically significant in distinguishing the disease from no-disease. These types of studies require genotyping DNA sequence of individuals in both case and control samples. The current technologies of genotyping individual DNA sequence only genotype a limited DNA sequence of each individual in the study. As a result, a large fraction of SNPs are not genotyped. Various imputation methods based on individual level data have been developed and successfully implemented to impute the missing SNPs. These approaches can broadly fit into two categories. One is the Hidden Markov Model (HMM) based approach. Some methods in this category are: IMPUTE [1,2], MACH [3,4], BEAGLE [5], and fastPhase/BIMBAM [6]. The other is tag SNP based approach. Some methods in this category are: TUNA [7], WHAP [8], and SNPMstat [9]. Since these methods are based on individual level DNA sequence, they are time consuming and costly. Furthermore, the need of individual level data in the imputation process results in the exclusion of the studies with only summary data in imputation-based meta-analysis. Thus, it is desirable to develop methods for imputing the untyped SNPs from the summary level data.

The purpose of imputation is to increase the sample size and the coverage of SNPs, which in turn increases the power of the association test. However, imputation at individual level is time consuming and costly. Alternative approaches could be to impute the allele frequencies or p-values directly. By imputing the p-values directly would be the most efficient with the least cost. However, the p-value is associated with the association test. One can apply different association tests while each test may result in different p-values. Imputing p-values directly will require a pre-determined association test. Therefore, imputing allele frequencies seems to be a better alternative in terms of efficiency, cost and allowing for applying different association tests to identify the significant SNPs.

Gautam [10] developed a new method, namely the Minimum Deviation of Conditional Probability (MiDCoP), which aims at imputing the allele frequencies of the missing SNPs using the allele frequencies of neighboring SNPs without using the individual level SNP information. The advantages of this new method include (1) it does not require individual level genotype data. Thus it is much more computationally efficient, and (2) it can be applied to the studies where only the summary level data such as allele frequencies of SNPs are available. However, it is essential that the method has to perform properly in order to be a viable approach for practical use. Gautam [10] performed various evaluations regarding the estimating accuracy of missing SNPs, and concluded that the correlation between the actual and imputed allele frequencies is higher than .9925 in the study. In this article, we will perform further evaluation of the method by evaluating the performance of the association tests using the case-control data of Genotype Association Information Network (GAIN) Schizophrenia study in European American Population. Section 2 gives a brief description of the MiDCoP method. Section 3 briefly describes the GAIN data and the general algorithm implemented for our evaluation. Section 4 summarizes the performances of the association statistics based on imputed and actual allele frequencies of SNPs using the GAIN data. Section 5 compares the association test results using different reference data sets. Section 6 gives a brief summary and conclusion.

2   The MiDCoP Method

The idea behind the Mimimum Deviation of Conditional Probability method (MiDCoP) is to impute the allele frequencies of untyped SNPs in the study sample by utilizing the allele frequencies of neighboring SNPs and haplotype frequencies from an external reference set such as the HapMap reference set (The International HapMap Project [11]). The best pair of the neighboring SNPs is determined by maximizing certain multilocus information score (MIS). Gautam [10] proposed five different MISs. In this article, we will adopt the best MIS recommended in [10], namely, the Mutual Information Ratio (MIR, [12]). The algorithm of the (MiDCoP) derived by Gautam [10] consists of the following three steps:

1) SNPs Selection: Identify a set of flanking SNPs in the neighborhood of the untyped SNP X that maximize MIR based on reference set. Let L = {L$_1$, L$_2$, ..., L$_u$} be the sequence of SNPs common to both reference set and sample set in the neighborhood of X, and are in linkage disequilibrium with X. Our goal is to obtain a pair {A, B} $\subseteq$ L such that the obtained MIR between {A, B} and {A, X, B} in the reference set is maximized for the fixed SNP X. Here, the order of SNPs {A, B} does not need to be in the sequential order based on their base pair position.

2) Haplotype Frequency Estimation: Once the optimal pair {A,B} is determined from step 1, this step estimates the haplotype frequency for the pair {A, B} in the sample.

3) Allele Frequency Estimation: The allele frequency of untyped SNP X in the sample is estimated as the weighted sum of the haplotype frequency estimated in the step 2.

The Mutual Information Ratio (MIR) is defined in the following. Let S = {S$_1$, S$_2$, ..., S$_n$} and T = {T$_1$, T$_2$, ..., T$_m$} be two disjoint sets of n and m (bi-allelic) SNPs with the population haplotype frequencies given by the vectors $\varphi = (\varphi_1, \varphi_2, ..., \varphi_s)$ and $\theta = (\theta_1, \theta_2, ..., \theta_t)$, respectively. The unknown parameters

φ and θ, are estimated by $\hat{\varphi}$ and $\hat{\theta}$ through the reference sets in our study. The Shannon entropies of the discrete random variables S and T are given, respectively, by

$$e_{s} = -\sum_{j=1} \varphi_{j} \log \varphi_{j} \text{ and } e_{T} = -\sum_{i=1} \theta_{i} \log \theta_{i}.$$

If e_{ST} is the entropy of the joint distribution, the mutual information between S and T is defined as $MI(S,T) = e_S + e_T - e_{ST}$ which is bounded by $\min(e_s, e_T)$. The MIR is defined as the normalized mutual information [12] and is given a $MIR(S,T) = MI(S,T) / \min(e_s, e_T)$. The MIR measure can be considered as the shared information between the two sets of haplotypes. It is symmetric to both S and T, i.e., $MIR(S,T) = MIR(T,S)$. For a missing SNP X, the MIR is computed between the set $S=\{A,B\}$ and the missing SNP $\{X\}$. The flanking SNPs selection step is to choose the pair $\{A, B\}$ which estimates the haplotype distribution for the set $\{A, X, B\}$ with minimal loss of information.

3   Data Source

In order to evaluate and compare the performance of the MiDCoP method under different scenarios, we analyze the case-control GWAS data on the study of Genotype Association Information Network (GAIN) Schizophrenia in European American Population from the Database of Genotype and phenotype (dbGap) (from: http://www.ncbi.nlm.nih.gov/sites/entrez?db=gap, [13]). This data set (dbGap analysis accession: phs000021.v3.p2) consists of 1,351 Schizophrenia cases and 1,378 controls of European American Population genotyped by using the Affy 6.0 [13]. The HapMap III (The International HapMap Consortium, [11]) phase-known data for CEU population (U.S. residents of northern and western European ancestry (CEU)) are used as the reference set. The reference set has 234 counts of haplotypes.

4   Overall Performance of MiDCoP Method Using the GAIN Data Set

Our purpose is to investigate the accuracy of the association test when the allele frequencies in case and control are imputed using the MiDCoP method. The allele-based tests [14] are applied for this purpose. For comparison, we assume that each SNP is missing, estimate the allele frequency of the 'missing' SNP, and compute the allele-based association test using the actual and imputed allele frequency for each SNP. The test statistics and the corresponding p-values are compared. The evaluation is carried out on several regions of genomes using the GAIN Schizophrenia GWAS data. The procedure of the evaluation is described in the following.

1) Randomly select a region of 3,000 SNPs from each of the 22 chromosomes of the GAIN data and drop the SNPs with minor allele frequency (MAF) ≤ 0.05 in the reference set.
2) Impute the allele frequency for each SNP using the MiDCoP method in both case and control data by assuming it is missing. The MIR multilocus information measure is used to select the optimal pair of SNPs for imputing the missing SNP.
3) Compute the association test statistic and the corresponding p-value between case and control for each SNP based on the imputed allele frequencies. Note that the p-values are transformed to the negative logarithm (base 10).
4) Repeat Step 3 for each SNP based on the actual allele frequencies between case and control.
5) A simple linear regression is computed to fit the p-values from imputed allele frequencies (Y) with the p-values from the actual allele frequencies (X). If the imputation is perfect, then, the intercept is zero, slope is one and the coefficient of determination (R^2) is one.

Among the 22 chromosomes studied, the R^2 ranges from 0.62 to 0.79 (or in terms of correlation, 0.787 to 0.889). The results are further summarized based on the categorized levels of pairwise linkage disequilibrium (LD) of SNPs. SNPs with maximum pairwise LD ≥ 0.75 is classified as High LD SNPs. Similarly, other labels are defined as: Moderate, Low, and Weak LD if the maximum pairwise LD is in the range of [0.5, 0.75), [0.25, 0.5), and [0, 0.25), respectively. Figure 1 displays the scatter plots of $-\log_{10}$(p-values) between imputed (Y) and actual (X) SNPs for all 22 chromosomes.

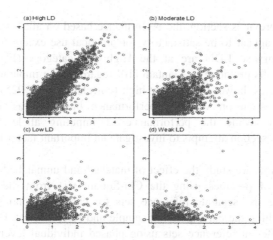

Fig. 1 Scatter plots of $-\log_{10}$(p-values) of all 22 chromosomes between imputed (Y) and actual (X) SNPs

The comparison of R^2 and simple linear regression for Chromosome 1 is summarized in Table 1. A finer categorization of the LD than that used for the scatter plots in Figure 1 is used in Table 1. Results for other chromosomes are similar.

Table 1 Comparison of p-values between imputed (Y) and actual (X) allele frequencies of 3000 SNPs from Chromosome 1 of the GAIN data

LD Group	R^2	Intercept	Slope
LD=1	0.9398	0.0088	1.0064
LD[0.9-1.0)	0.7504	0.0441	0.9305
LD[0.8-0.9)	0.6832	0.1163	0.9032
LD[0.7-0.8)	0.4438	0.1938	0.8084
LD[0.6-0.7)	0.4882	0.0979	0.8310
LD[0.5-0.6)	0.7264	0.0784	1.1980
LD[0,0.5)	0.1834	0.2575	0.6206
All	0.7818	0.0637	0.9555

Both scatter plots of all 22 chromosomes and results in Table 1 for Chromosome 1 are consistent and indicate that the p-values based on imputed allele frequencies are biased toward underestimation. The performance seems to be appropriate for situations when the neighboring LD values are high and is getting worse as the LD values decreases.

5   Performance of MiDCoP Using Different Reference Sets

As shown in previous section, the performance based on allele-based association tests does not appear to be satisfactory for practical use except for the situations where the pairwise LD's are at the High LD category (Figure 1(a).) The simulation results provided by Gautam [10] regarding the imputation accuracy of allele frequencies, however, are very high (correlation > 0.9925). This raises an interesting question on why the performance based on association tests is unsatisfactorily; while the performance of imputing allele frequencies are excellent. This section attempts to investigate an important factor that leads to this discrepancy.

In this section, we study the effect of matched and unmatched reference set to the accuracy of p-values using the Chi-Square test from the imputed allele frequency. We use separate reference sets for case and control with different levels of matching the corresponding sample and reference set. First, we need to create two different reference sets using phased individual level genotype data. One reference set is the 'unmatched' reference set, that is, the individuals in the reference set are not overlapping with those in the sample data that will be imputed. The other is the 'matched' reference set. Instead of creating a 100%

matched reference set with the case and control data, respectively, we create a reference set with 2/3 of individuals overlapping as a more realistic situation. We could have chosen 3/4 or 4/5 overlapping. However, 2/3 seems to more realistic and practical from the real data we have observed. The data set and procedure of creating the reference sets is described in the following.

The phased individual level genotype data of 1,150 cases and 1,378 controls from the GAIN Schizophrenia is used. A region of 50 MB-60 MB is selected from the Chromosome 2 for the study. We choose Chromosome 2 for the reason that the overall performance of association tests is the third worst among all 22 chromosomes (only better than Chromosome 9 and 22), and that the number of SNPs available for our study is the second largest among all chromosomes (second to Chromosome 1).

Two random samples of equal size are generated from the case data. Two different scenarios for generating the random samples are studied.

- Scenario One: There is no individual overlap between the two random samples.
- Scenario Two: There is exactly 2/3 of individual overlaps between the two random samples.

In the first scenario, each cohort is randomly split into two halves so that there is no overlap between the two subsets. In the second scenario, two subsets are generated in such way that 50% of the individuals randomly selected from the original data belonging to both subsets and the remaining 50% individuals are split randomly into each of the two subsets. We use one of the subset as the reference set and the other as the study sample.

In order to imitate the imputation process of MiDCoP, we compute the allele frequencies for the SNPs in the study sample. Then, the allele frequencies of each of the SNPs in the study sample are imputed using the MiDCoP approach by assuming they are missing. This process is carried out for both case and control data. The p-values of the Chi-square tests from the actual and imputed allele frequencies are compared using the linear regression for SNPs in different pairwise LD groups. The process of selecting the subsets and computing the association test are repeated for fifty times (number of simulations) under both scenarios. The coefficient of determination (R^2), intercept, and slope of the linear fit are obtained from each of the 50 simulations. Summary statistics (mean, standard deviation, maximum, and minimum) are computed from the regression statistics under both scenarios. Table 2 and Table 3 summarize the results for the first and second scenario, respectively.

Comparison of results from Table 2 and Table 3 suggests that matching the sample with the reference set leads to higher imputation accuracy. Note that the conditional probability, $P(X|A-B)$, used in the MiDCoP approach to impute the allele frequency in both scenarios are computed from different reference sets for case and control.

Table 2 Summary of linear regression fit of p-values between imputed (Y) and actual (X) SNPs for non-matched scenario

LD group	Reg. Statistics	Mean	Sta. Dev	Min	Max
1	R^2	0.8401	0.0233	0.7895	0.8803
	Intercept	0.0338	0.0057	0.0223	0.0474
	Slope	0.9204	0.0176	0.8831	0.9535
[0.9-1.0)	R^2	0.7228	0.0451	0.6159	0.8143
	Intercept	0.0625	0.0126	0.0367	0.0912
	Slope	0.8499	0.0374	0.7672	0.9281
[0.8-0.9)	R^2	0.4878	0.061	0.3286	0.6091
	Intercept	0.1182	0.0203	0.0713	0.1782
	Slope	0.6975	0.0568	0.5094	0.7935
[0.7-0.8)	R^2	0.2946	0.0756	0.1449	0.4358
	Intercept	0.1782	0.0331	0.1164	0.2688
	Slope	0.5517	0.083	0.3683	0.7181
[0.6-0.7)	R^2	0.2057	0.0836	0.0288	0.4487
	Intercept	0.2332	0.0428	0.15	0.3481
	Slope	0.4542	0.1081	0.1729	0.7369
[0.5-0.6)	R^2	0.0707	0.0487	-0.0042	0.2084
	Intercept	0.2853	0.0417	0.1992	0.3761
	Slope	0.2748	0.1132	0.0561	0.6046
[0,0.5)	R^2	0.0076	0.0121	-0.0029	0.0411
	Intercept	0.3486	0.0278	0.299	0.4203
	Slope	0.0837	0.0683	-0.0139	0.2604
Over All	R^2	0.5179	0.0421	0.444	0.5968
	Intercept	0.1126	0.0101	0.094	0.1384
	Slope	0.7236	0.0334	0.6454	0.7893

The gain in imputation accuracy under the second scenario, when there is a 2/3 match between the reference set and sample, also suggests that the conditional probability P(X|A-B) in the Step 2 of the MiDCoP method should not be fixed but should be adjusted to match with the sample data to be imputed.

Table 3 Summary of linear regression fit of p-values between imputed (Y) and actual (X) SNPs for 2/3 matched scenario

LD group	Reg. Statistics	Mean	Sta. Dev	Min	Max
1	R^2	0.9427	0.0077	0.9236	0.9558
	Intercept	0.0119	0.0031	0.0037	0.0178
	Slope	0.9713	0.0090	0.9515	0.9903
[0.9-1.0)	R^2	0.8893	0.0178	0.8522	0.9269
	Intercept	0.0237	0.0071	0.0103	0.0393
	Slope	0.9437	0.0203	0.8954	0.9832
[0.8-0.9)	R^2	0.7904	0.0319	0.7159	0.8432
	Intercept	0.0462	0.0113	0.0213	0.0736
	Slope	0.8859	0.0332	0.7836	0.9603
[0.7-0.8)	R^2	0.6828	0.0408	0.6077	0.8001
	Intercept	0.0652	0.0147	0.0365	0.0978
	Slope	0.8289	0.0438	0.7103	0.9372
[0.6-0.7)	R^2	0.6452	0.0607	0.4916	0.7944
	Intercept	0.0872	0.0242	0.0368	0.1465
	Slope	0.8040	0.0611	0.6443	0.9406
[0.5-0.6)	R^2	0.4555	0.0772	0.2568	0.6204
	Intercept	0.1210	0.0311	0.0566	0.2116
	Slope	0.6718	0.0940	0.4287	0.8971
[0,0.5)	R^2	0.3084	0.0581	0.1825	0.4124
	Intercept	0.1656	0.0225	0.1115	0.2160
	Slope	0.5673	0.0653	0.4299	0.7024
Over All	R^2	0.7902	0.0144	0.7548	0.8216
	Intercept	0.1656	0.0225	0.1115	0.2160
	Slope	0.8904	0.0137	0.8587	0.9166

6 Summaries and Conclusion

GWAS studies have been an important area of studies to investigate the associations of SNPs between diseases and genetic markers. Due to technological limitation, a large fraction of SNPs in the genome are usually not genotyped. In order to increase the opportunity of identifying the SNPs with potential high association with diseases and increase the power of the association tests, several methods for imputing the missing SNPs at the individual level data have been developed and successfully applied in GWAS studies. However, due to computational intensity and cost, there is a need for developing methods to impute the summary data of SNPs such as allele frequencies or the p-values of the association tests directly. A new method namely MiDCoP method to impute the allele frequencies of missing SNPs was developed in [10]. This article investigates the performance of the MiDCoP method by using the case-control GWAS data on the study of Genotype Association Information Network (GAIN) Schizophrenia in European American Population. We first evaluate the association tests by comparing the corresponding p-values between imputed and actual allele frequencies of SNPs. The results appear that the MiDCoP performs adequately only when the LD values are high. The second evaluation is to compare the performance between 'matched' and 'unmatched' reference sets. The results indicate that the better the reference set matches the sample data, the better the performance of the method is.

Several questions remain unanswered in this article. First, in a practical problem, it is not easy to find the reference that 'matches' the sample data. In most practical situations, the imputation in both case and control is carried out with a single reference set. In our study, we use the reference set from the HapMap project. It is critical to identify a reference set from the population that matches the sample data as close as possible. Besides the HapMap project, one can also look for the reference set from the 1000 Genome project [14]. As the results indicate that there is an underestimate bias based on the MiDCoP method. Further research will be needed to develop a method to adjust the bias. One possible approach is to adjust the conditional probability, $P(X|A-B)$ from the reference set based on the bias adjustment of the allele frequencies of the flanking SNPs in case and control. Another further research is to compare the performance between the MiDCoP method and the existing individual level based imputation methods such as IMPUTE. The authors have made some progress in this research. More analysis and comparisons will be needed using existing GWAS data.

Acknowledgement. This research is partially supported by the internal grant from Central Michigan University. The GWAS data analyzed is the data set of GAIN Schizophrenia for European ancestry from Database of Genotypes and Phenotypes (dbGap) (Bethesda, MD: National Center for Biotechnology Information, National Library of Medicine. Available from: http://www.ncbi.nlm.nih.gov/sites/entrez?db=gap. dbGaP analysis accession: pha0002857.v1.p1) and the HapMap III (The International HapMap Project, 2010) reference panel of CEU population.

References

1. Marchini, J., Howie, B., Myers, S., McVean, G., Donnelly, P.: A new multipoint method for genome-wide association studies by imputation of genotypes. Nature Genetics 39, 906–913 (2007)
2. Howie, B., Donnelly, P., Marchini, J.: A flexible and accurate genotype imputation method for the next generation of genome-wide association studies. PLoS Genetics 5, e1000529 (2009)

3. Li, Y., Ding, J., Abecasis, G.R.: Mach 1.0: Rapid Haplotype Reconstruction and Missing Genotype Inference. The American Journal of Human Genetics 79, S2290 (2006)
4. Li, Y., Willer, C.J., Ding, J., Scheet, P., Abecasis, G.R.: MaCH: Using sequence and genotype data to estimate haplotypes and unobserved genotypes. Genetic Epidemiology 35, 816–834 (2010)
5. Browning, B., Browning, S.R.: A unified approach to genotype imputation and haplotype-phase inference for large data sets of trios and unrelated individuals. The American Journal of Human Genetics 84, 210–223 (2009)
6. Guan, Y., Stephens, M.: Practical Issues in Imputation-Based Association Mapping. PLoS Genetics 4(12), e1000279 (2008), doi:10.1371/journal.pgen.1000279
7. Nicolae, D.L.: Testing untyped alleles (TUNA)-applications to genome-wide association studies. Genetic Epidemiology 30, 718–727 (2006)
8. Zaitlen, N., Kang, H.M., Eskin, E., Halperin, E.: Leveraging the HapMap correlation structure in association studies. American Journal of Human Genetics 80, 683–691 (2007)
9. Lin, D.Y., Hu, Y., Huang, B.: Simple and efficient analysis of disease association with missing genotype data. The American Journal of Human Genetics 82, 444–452 (2008)
10. Gautam, Y.: A novel approach of imputing untypes SNP using the allele frequencies of neighboring SNPs. Unpublished dissertation, Central Michigan University, USA (2014)
11. The International HapMap Consortium: Integrating common and rare genetic variation in diverse human populations. Nature 467, 52–58 (2010)
12. Zhang, L., Liu, J., Deng, H.W.: A multilocus linkage disequilibrium measure based on mutual information theory and its applications. Genetica 137, 355–364 (2009)
13. Database of Genotype and phenotype (dbGap): Available at Bethesda (MD): National Center for Biotechnology Information, National Library of Medicine, http://www.ncbi.nlm.nih.gov/sites/entrez?db=gap
14. Zheng, G., Yang, Y., Zhu, X., Elston, R.C.: Analysis of Genetic Association Studies. Springer, New York (2012)
15. The 1000 Genomes Project Consortium: An integrated map of genetic variation from 1,092 human genomes. Nature 491, 56–65 (2012)

An Approach to Construct Weighted Minimum Spanning Tree in Wireless Sensor Networks

Soumya Saha and Lifford McLauchlan

Abstract. Topology control is critical to extend the lifetime of energy constrained Wireless Sensor Networks (WSNs). Topology control mechanism can be divided into two processes: topology construction and topology maintenance. During topology construction one creates a reduced topology to ensure network connectivity and coverage. In topology maintenance, one recreates or changes the reduced topology when the network is no longer optimal. In this research the authors concentrate on Minimum Spanning Tree (MST) which is a commonly seen problem during the design of a topology construction protocol for WSNs. As the amount of running time and messages successfully delivered are important metrics to measure the efficacy of distributed algorithms, much research to create simple, local and energy efficient algorithms for WSNs thereby creating sub optimal MSTs has been studied. In this research, two popular approaches are discussed to create a Spanning Tree in the WSNs- Random Nearest Neighbor Tree (Random NNT) and Euclidian Minimum Spanning Tree (Euclidian MST). Next, the authors propose a method which has the goals to balance the network load evenly among all of the nodes and increase the number of successful message deliveries to the sink. Finally a comparison between the three algorithms is conducted in the Matlab environment. Simulation results demonstrate significant improvement for both load balancing and number of message deliveries after implementation of the proposed algorithm.

Keywords: Topology construction protocol, Minimum Spanning Tree, Nearest Neighbor Tree, Load balancing, Simple Weighted Spanning Tree.

Soumya Saha · Lifford McLauchlan
Department of Electrical Engineering and Computer Science, Texas A&M University,
Kingsville, TX, 78363 USA
e-mail: jishumail@gmail.com,
       lifford.mclauchlan@tamuk.edu

© Springer International Publishing Switzerland 2015   69
R. Lee (ed.), *SNPD*,
Studies in Computational Intelligence 569, DOI: 10.1007/978-3-319-10389-1_6

1   Introduction

Wireless Sensor Networks (WSNs) have become more prevalent due to their relatively cheap cost and increased sensor node processing capability. WSNs will likely continue to have important roles in both civil and military applications. The nodes in a WSN are small devices which possess sensors, communication capabilities, processing units and memories and power sources. Progress in decreasing the size of the battery has not kept up with the rate at which electronic circuits have become smaller, and thus, the power source has become a very critical issue while developing WSNs and associated algorithms for constructing and maintaining any wireless sensor network. In many applications, the sensor nodes are deployed randomly in the application's region of interest. The network is formed by self-configuration of the nodes; a simple and local distributed algorithm has the capability for the nodes to configure themselves in an energy efficient manner. Fig. 1 depicts a set of randomly deployed sensor nodes.

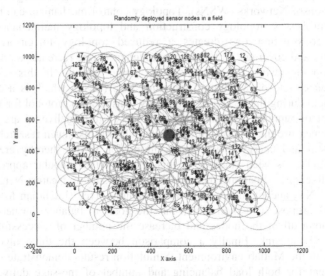

Fig. 1 240 sensors are deployed in uniformly random manner in a 1000m x 1000m area of interest. A sink node is depicted by the large solid red circle while the other nodes are represented by the small blue dots. Each node's maximum communication area is depicted by the green circle centered at the node and encircling it.

In the literature, Minimum Spanning Tree (MST) algorithm has been utilized to make the Connected Dominating Set of the nodes resulting in the communication backbone for the WSN [5]. Much research has been conducted on the weighted Spanning Tree, in which the nodes are chosen by their rank or weight. In this research, the authors discuss two basic approaches to form the wireless sensor networks to make the MST: Random Nearest Neighbor Tree (Random NNT) and Euclidian MST (also known as Geometric weighted MST). The authors have developed an energy efficient weighted MST algorithm- Simple Weighted Spanning Tree (SWST), and analyzed and compared it with the two above mentioned algorithms.

Dynamic Global Topology Recreation (DGTRec) was implemented as the topology maintenance protocol that will recreate the reduced topology at predefined time intervals. The simulation results show improvements in network load balancing after utilization of the proposed SWST algorithm. The SWST algorithm has many advantages such as being easily scalable and distributed in nature as well as exhibiting increased load balancing and simplicity. The rest of the sections are organized as follows: Related Work is discussed in section 2, section 3 proposes the Simple Weighted Spanning Tree (SWST) algorithm, section 4 includes the Performance Evaluation and the last section is the Conclusion.

2 Related Work

Distributed algorithms for topology construction (TC) in WSNs utilize and require significantly less hop information than centralized algorithms. There are two commonly employed methods for distributed algorithms- cluster head technique and connected dominating set (CDS) technique. For the latter approach, determination of minimum connected dominating set (MCDS) is one of the most important tasks to build the communication backbone [6]. Much research has been published in the literature that addresses this problem [7-14]. Many proposed algorithms construct the topology by first creating a preliminary CDS and adding or removing nodes to or from the set to create an approximate MCDS [5, 15]. Two energy efficient CDS based topology construction protocols- CDS-Rule-K and Energy Efficient CDS (EECDS) were discussed in [11, 14]. Also, Euclidian Minimum Spanning Tree (Euclidian MST) and Random Nearest Neighbor Tree (Random NNT) was proposed in the literature to construct low-cost spanning tree in WSNs [1].

A. Random Nearest Neighbor Tree: The Nearest Neighbor Tree algorithms avert many of the issues resulting from the overhead produced by the Minimum Spanning Tree by picking a unique rank and making a node connect with a node of higher rank [1]. In fact, it is in many ways similar to an approximation algorithm for solving the Traveling Salesman problem in [2, 3]. The Random NNT is an effective method to create a low cost spanning tree. In this algorithm, nodes pick a random rank between 0 and 1 and connect to the closest node of higher rank [1]. To start the topology construction, a predefined node begins broadcasting messages. All nodes that receive a broadcasted message will send an acknowledgment message to the sender. The sender node will connect to recipient node since it possesses higher rank than the sender. As the next step, the recipient node broadcast messages. This is then repeated in the network at every node thereby building a spanning tree. The description for the following algorithm is from [16].

Basic algorithm for Random NNT running at each node u:
- d=maximum possible distance between any two nodes
- when a message is broadcasted, the recipient is not specified

$i \leftarrow 1$

Repeat

    Set transmission radius (power level): $r_i \leftarrow 2^i/\sqrt{n}$

    If $r_i > d$, set $r_i \leftarrow d$

        Broadcast (request, Id of u ID(u), rank info p(u) which is a random number)

        $i \leftarrow i+1$ until receipt of an available message or $r_i=d$

Upon receipt of an available message from any node v Do

    If rank(v) > rank(u)

        Set transmission radius $\leftarrow$ distance(u,v)

        Send available message to v

Upon receipt of all available messages

    Select the nearest node v from the senders

    Send (connect u, v) to v

B. Euclidian Minimum Spanning Tree: The Euclidian Minimum Spanning Tree, also known by Geometric weighted Minimum Spanning Tree, is an Euclidian distance weighted MST. This algorithm will attempt to minimize the power consumption during message transmission from the transceiver. Since energy dissipated to transmit a message is proportional to the square of the transmission distance, an algorithm that minimizes the transmission distance is a logical method to address reducing the energy used for transmission. The description for the following algorithm is modified from the description in [16].

Basic algorithm for Euclidian MST running at each node u:
- d=maximum possible distance between any two nodes
- when a message is broadcasted, the recipient is not specified

$i \leftarrow 1$

Repeat

    Set transmission radius (power level): $r_i \leftarrow 2^i/\sqrt{n}$

    If $r_i > d$, set $r_i \leftarrow d$

        Broadcast (request, Id of u ID(u), power level to help calculate the RSSI)

        $i \leftarrow i+1$

Upon receipt of all available messages

    Select the nearest node v from the senders

    Set transmission radius $\leftarrow$ distance(u, v)

    Send available message to v

    Send (connect u, v) to v

   The next section will discuss the Simple Weighted Spanning Tree (SWST) algorithm.

3   The SWST Algorithm

The proposed Simple Weighted Spanning Tree (SWST) algorithm builds a weighted Minimum Spanning Tree in which the weight is one of the following: parent node or children node energy level; or the Euclidian distance between parent and children nodes. Due to parent nodes exhibiting a closer proximity to the sink node (lower hop numbers), parent nodes will transmit messages more often on average during a sensing event resulting in the parent nodes more likely depleting their energy resources before children nodes (higher hop numbers) will.

   The proposed SWST algorithm mainly focuses on load balancing among parent nodes. In this SWST protocol, children nodes choose parent nodes exhibiting the highest energy level. For this reason, parent nodes will enable the network to run for a more extended period of time until the topology maintenance (TM) protocol starts. Each node has a threshold energy level associated with it; if the node energy falls under this threshold, the node tends to utilize Euclidian MST protocol in order to increase its own node life. The energy consumed to transmit decreases due to the reduced transmission range of the transceiver. After a node chooses its final parent, its transmission range will be set to the Euclidean distance between the final parent node and itself.

   However, if the energy level of the child node is more than the maximum energy level of the parent nodes for the child node, energy weighted MST will be utilized instead of Euclidian MST protocol. This is due to the fact that it is better to have more energy in the parent nodes than the children nodes when using the SWST algorithm. If children nodes exhibit higher energy levels, the parent nodes will deplete energy sooner and produce an unconnected link in the communication backbone. In that case, children nodes will not be able to communicate with the sink node and thus, a child node will attempt to choose the strongest node even if that depletes its own resources.

   Sink node: A sink node has significantly more energy and longer transmission range than the ordinary sensor nodes. In the TC phase, sink nodes set their transmission range equal to the range of ordinary nodes since child nodes that exhibit a shorter transmission range would be incapable of transmitting messages directly to the sink node if the distance from the child node to the sink node is further than the child node's own transmission range. During topology maintenance, the sink node switches its transmission range to the highest level to initiate the maintenance protocol in all ordinary nodes within the maximum transmission distance of the sink node. This description assumes all nodes are within the sink node's maximum transmission range.

   Active nodes: Active nodes possess energy levels sufficient to perform sensing and communication. These nodes actively listen for message delivery request and sense the environment for an event of interest. They deliver messages for events of interest or as requested to its default gateway. Active nodes are default gateways for other active or sleeping nodes. Active nodes together form the virtual communication backbone of the network.

Sleeping nodes: In order to conserve the energy of the nodes, the nodes can turn their radio off instead of actively listening for messages. In addition, the nodes if they sense any events by themselves will transmit the message concerning the sensing event; sleeping nodes will not relay messages for other nodes.

Comatose nodes: Comatose nodes may possess sufficient energy to sense and transmit messages, but due to their location, they are not within the communication range of any active or sleeping nodes. Since they cannot deliver a message to the sink due to their unconnected status, they will generally not contribute sensing or transmitting capability to the network.

Dead nodes: In a WSN, nodes have a finite amount of energy. As the node's energy is depleted the node will become incapable of sensing events or transmitting messages. Such nodes are denoted as dead nodes.

Successful Event: A successful event occurs during the operation of the WSN when a message for a sensing event as determined by either a sleeping or an active node is delivered and received by the sink node. From the sink node, the message is transmitted to the Base Station (BS). In this paper, the authors maintain that the number of successful events is very critical to successful WSN operation since the main goal of a WSN is to sense events and send the corresponding messages to the Base Station.

Energy Model: The energy model assumed in this research was adopted from [4]. The amount of energy consumed in transmission and reception of data can modeled as:

$$E_t(k,r) = kE_{elec} + k\varepsilon r^2 \tag{1}$$

$$E_r(k) = kE_{elec} \tag{2}$$

where,

$E_t(k,r)$ = energy required for transmission
$E_r(k)$ = energy required for listening
k = the number of bits in each message
E_{elec} = the energy needed to run the transceiver
ε = energy consumption of the amplifier per unit area
r = distance between parent and child nodes

During testing of the proposed algorithm, the simulations assumed same data size for each message. In addition the transmission energy dissipated is proportional to the distance between parent and child nodes as seen in equation 1 and for listening the energy is fixed as in equation 2. As stated before, active nodes will constantly listen for message delivery request. When a message delivery request is received, an active node will transmit a message. In addition, if an event is sensed by an active or sleeping node, it will send a message. Fig. 2 depicts the communication backbone after a random WSN deployment.

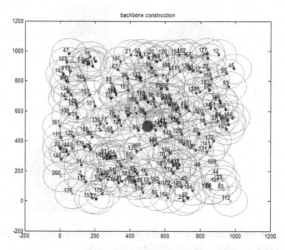

Fig. 2 Communication backbone construction after the random deployment of 240 nodes in a 1000m x 1000m field: small red, blue and yellow dots represent active, sleeping and comatose nodes respectively. The large solid red circle is the sink node. The green circles around each node represent maximum communication area of the respective nodes.

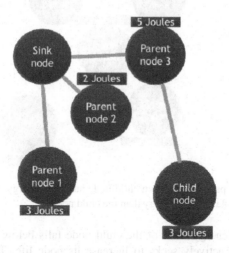

Fig. 3 Selection of parent node when child node has more energy than the threshold value

The proposed algorithm is illustrated in Figs. 3, 4 and 5. In Fig. 3, the child node has more energy than the threshold energy level (1.5 Joules) and it has three options- parent 1, parent 2 and parent 3 to choose as its final parent. As parent 3 has the highest level of energy, the child node will choose it even though the distance between parent 3 and the child node is greater than that of parent 2 and the child node. In this case, the child node will increase its transmission power consumption to use the parent node with the highest energy as the default gateway and thus balance the load among the parent nodes.

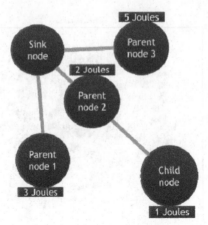

Fig. 4 Selection of parent node when child node has more energy than the threshold value and all parent nodes has more energy than the child node

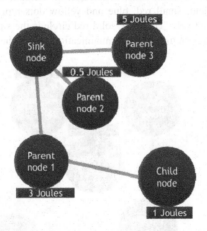

Fig. 5 Selection of parent node when child node has more energy than the threshold value and some parent node has less energy than the child node

In Fig. 4, the energy level of the child node falls below the threshold energy level and thus it actively seeks to increase its node life. Thus the node adopts Euclidian MST approach and finds the nearest parent node. In this way, it saves its energy by reducing its transmission range.

However, in Fig. 5, the energy level of the child node is below the threshold energy level and the nearest parent node's energy is less than that of the child node. If the child node chooses parent 2 to reduce its transmission power consumption, the parent node will die sooner than child node and the child node will not be able to deliver any message to the sink node. In that case, the child node will not enjoy any benefit saving its energy because it can no longer fulfill its basic duties- to report events and deliver messages to the sink node. Therefore, it will choose the nearest parent node which has more energy than itself, and in this case, parent 1. The following algorithm is also discussed in [17].

Proposed SWST algorithm for each node u:
E(u)=remaining energy,
status(u)=either active, sleeping or comatose,
D(u)=set of distances between the current node and the sender nodes,
PE(u)=set of energy levels of the parent nodes,
P(u)=set of parent node IDs,
FP(u)=final parent node ID,
tr(u)=transmission range,
TEL=Threshold Energy Level.

In this study, one sink node is assumed. The sink node starts the TC by broadcasting a Hello message. Hello messages include three items which are the sender node ID, sender node's energy level and sender node's transmission power level in order to enable the determination of the RSSI (if each node's maximum transmission range is fixed, then one can remove this portion of the message and decrease the message length thereby saving energy).

The following algorithm description is also found in [17].

1) Initialization:
P(u)=Ø, FP(u)= Ø, PE(u)= Ø, D(u)= Ø, status(u)=comatose, tr(u)=maximum transmission range.

2) Parent Discovery:
Starts Parent Discovery timeout.
Starts Hello message listening timeout.
Hello message listening timeout expires.
If does not receive any Hello message
    status(u)=comatose;-
Else status(u)=active;
    update P(u): upon receiving Hello message from a sender node v, P(u)=P(u) U {v};
    update PE(u): PE(u)=PE(u) U {E(v)};
    update D(u): D(u)=D(u) U {distance(u, v)};
End if.
Parent Discovery timeout expires.
Broadcast Hello message to the children nodes.

3) Parent Selection:
Starts Parent Selection timeout.
Calculate max(PE(u))
If max(PE(u)) > E(u) && E(u) < TEL

Calculate min(D(u))
    Select the corresponding parent node from D(u) as
    FP(u):FP(u)=P(u){index(min(D(u)))};
Update tr(u) using RSSI: tr(u)=min(D(u));
Else
    Select the corresponding parent node from PE(u) as
    FP(u):FP(u)=P(u){index(max(PE(u)))};
Update tr(u) using RSSI:
tr(u)=D(u){index(max(PE(u)))};
End if.
Sends unicast parent recognition message to parent node.
    Parent Selection timeout expires.

4) *Set status:*
If status(u)=active
    If receives any parent recognition message
        status(u) = active;
    Else
        If does not receive parent recognition message
        status(u) = sleeping;
        End if.
    End if.
Else status(u) = comatose;
End if.

The proposed algorithm has the following advantages:

1) The SWST algorithm is scalable since it is a distributed algorithm; it does not require information on the overall network status and control.
2) Localization mechanisms such as GPS are not necessary.
3) The algorithm aims to balance the network load and energy. This is accomplished by saving energy by both parent and children nodes.
4) The algorithm is simple; during MST creation, only small amounts of running times are utilized and messages are exchanged which reduces energy consumption.
5) Message collisions are decreased significantly since the number of active nodes in the topology is fewer.

Dynamic Global Topology Recreation (DGTRec) was implemented as the topology maintenance protocol that will recreate the reduced topology at predefined time intervals. The simulation results show improvements in network load balancing after utilization of the proposed SWST algorithm.

4   Performance Evaluation

In this section, the authors simulated two basic approaches to form the wireless sensor networks to make the MST: Random Nearest Neighbor Tree (Random NNT) and Euclidian MST (also known as Geometric weighted MST) and compared both of them with the proposed Simple Weighted Spanning Tree (SWST) algorithm. In the simulations, the following was assumed:

1) Nodes are placed randomly in two dimensional Euclidian space.
2) Nodes exhibit perfect disk coverage.
3) Nodes exhibit same maximum transmission range and same sensing range.
4) All nodes are within the sink node's maximum transmission range.
5) Nodes do not possess information concerning their position, orientation and neighbors.
6) For simplicity, packets are assumed to not be lost in the Data Link layer.
7) Distance between two nodes is determined using Received Signal Strength Indicator (RSSI).
8) Network duty cycle = 100%.
9) Two or more simultaneous events in the network cannot occur, and rate of event occurrence is one event per second.

For each experiment, nodes were randomly deployed along with event coordinates and sequences. This information was then utilized with the other two algorithms tested to ensure a more accurate comparison and evaluation of the three algorithms. The simulation parameters are found in Table 1.

Table 1 Simulation Parameters

Parameters	Value
Deployment area	1000m x 1000m
Event number	1000
Time of simulation	1000 seconds
Number of events per unit time	1 event/second
Initial energy of the sink node	Infinity
Maximum transmission range of the sink	Infinity
Topology Maintenance after the amount of time	200 seconds
Energy consumed to transmit message for maximum range	0.1 Joule
Energy consumed for listening in 1 second	0.01 Joule
Threshold Energy Level	1.5 Joule

A. Experiment 1: Changing the node degree

In this experiment, maximum transmission ranges were changed to evaluate node degree. From Fig. 6 one can see that the proposed SWST algorithm outperforms the other two tested protocols in delivering the event sensing message to the sink node. When the node degree increases, children nodes are able to choose a strong parent enabling the system to perform better. As the number of successful events increases with the node degree, the energy required for message delivery will also increase. Therefore, the SWST algorithm will possess a high total number of dead nodes and a low total amount of remaining energy for a higher node degree. As for any node degree, SWST algorithm more successfully delivers messages to the sink, but in so doing will utilize more energy resulting in more dead nodes compared to the other two tested algorithms. As can be observed from Fig. 6, as the transmission range increases, the SWST algorithm outperforms the other protocols when comparing the number of successful events.

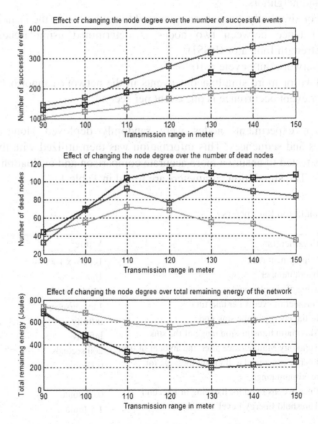

Fig. 6 Effect of various transmission ranges over different algorithms. Green, blue and red line represents Random NNT, Euclidian MST and the proposed SWST algorithm respectively.

B. Experiment 2: Changing the initial energy of the nodes

This experiment was conducted to observe the effect of changing the initial energy of each node over the number of successful events, total number of dead nodes and total amount of remaining energy. From Fig. 7 one can observe that, the number of successful events and the total amount of remaining energy increases, and the total number of dead nodes decreases with the increment of initial energy, as expected. From Fig. 7 one can see that, the proposed SWST algorithm consistently performs better than the other algorithms in delivering the event messages.

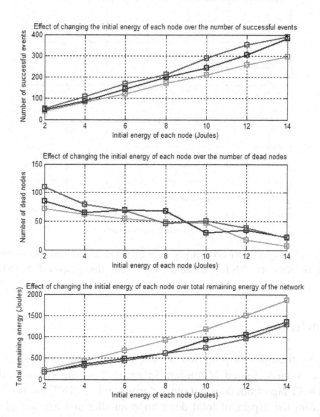

Fig. 7 Effect of various initial energy over different algorithms. Green, blue and red line represents Random NNT, Euclidian MST and the proposed SWST algorithm respectively.

C. Experiment 3: Changing the network density

Changing the network density by deploying more nodes is an alternative way to increase the node degree of each node in the network. Consequently, each node has more options to choose its parent node and hence increases the longevity of the system and number of successful events. However, total number of dead nodes is high and the amount of remaining energy of the system is low for a densely deployed network. This is shown in Fig. 8.

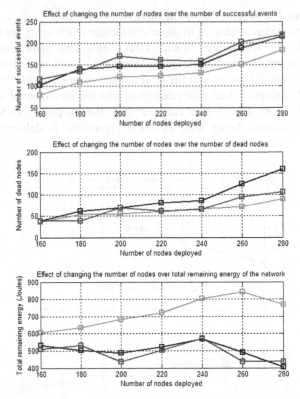

Fig. 8 Effect of changing network density over different algorithms. Green, blue and red line represents Random NNT, Euclidian MST and the proposed SWST algorithm respectively.

5   Conclusion

In this paper, the authors proposed an energy balanced topology construction protocol based on weighted Minimum Spanning Tree algorithm. Although the proposed SWST algorithm does not outperform Random NNT and Euclidian MST algorithm from the aspect of total dead node number and amount of remaining energy of the system, it surely performs best among the three in delivering the event sensing messages to the sink and consequently to the base station, which is the foremost important task of a sensor network. Future work will entail further testing of the algorithm against more energy balancing algorithms for WSNs. The model will be extended to have the WSN deployed in a three dimensional space to investigate applications in a more realistic environment that can include large structures inhibiting wireless sensor node transmissions.

References

1. Khan, M., Pandurangan, G.: Distributed Algorithms for Constructing Approximate Minimum Spanning Trees in Wireless Sensor Networks. IEEE Transactions on Parallel and Distributed Systems 20(1), 124–139 (2009)
2. Cohen, L., Avrahami-Bakish, G., et al.: Real-time data mining of non-stationary data streams from sensor networks. Information Fusion 9(3), 344–353 (2008)
3. Wang, X., Ma, J., Wang, S., Bi, D.: Time Series Forecasting Energy-efficient Organization of Wireless Sensor Networks. IEEE Sensors Journal 7(1), 1766–1792 (2007)
4. Kui, X., Sheng, Y., Du, H., Liang, J.: Constructing a CDS-Based Network Backbone for Data Collection in Wireless Sensor Networks. International Journal of Distributed Sensor Networks 2013, Article ID 258081, 12 (2013), http://dx.doi.org/10.1155/2013/258081
5. Mario, P., Rojas, W.: Topology control in wireless sensor networks. PhD Dissertation, University of South Florida (2010)
6. Raei, H., Sarram, M., Adibniya, F., Tashtarian, F.: Optimal distributed algorithm for minimum connected dominating sets in Wireless Sensor Networks. In: IEEE Conference on Mobile Ad Hoc and Sensor Systems, pp. 695–700 (2008)
7. Butenko, S., Cheng, X., Oliveira, C., Pardalos, P.M.: A New Heuristic for the Minimum Connected Dominating Set Problem on Ad Hoc Wireless Networks, pp. 61–73. Kluwer Academic (2004)
8. Chen, B., Jamieson, K., Balakrishnan, H., Morris, R.: Span: An energy-efficient coordination algorithm for topology maintenance in ad hoc wireless networks. Wireless Networks 8(5), 481–494 (2002)
9. Guha, S., Khuller, S.: Approximation algorithms for connected dominating sets. Algorithmica 20(4), 374–387 (1998)
10. Kumar, V., Arunan, T., Balakrishnan, N.: E-span: Enhanced-span with directional antenna. In: Proceedings of IEEE Conference on Convergent Technologies for Asia-Pacific Region, vol. 2, pp. 675–679 (2002)
11. Wu, J., Cardei, M., Dai, F., Yang, S.: Extended dominating set and its applications in ad hoc networks using cooperative communication. IEEE Trans. on Parallel and Distributed Systems 17(8), 851–864 (2006)
12. Wu, J., Dai, F.: An extended localized algorithm for connected dominating set formation in ad hoc wireless networks. IEEE Transactions on Parallel and Distributed Systems 15(10), 908–920 (2004)
13. Wu, J., Li, H.: On calculating connected dominating set for efficient routing in ad hoc wireless networks. In: Proceedings of the 3rd ACM International Workshop on Discrete Algorithms and Methods for Mobile Computing and Communications, pp. 7–14 (1999)
14. Yuanyuan, Z., Jia, X., Yanxiang, H.: Energy efficient distributed connected dominating sets construction in wireless sensor networks. In: Proceeding of the 2006 ACM International Conference on Communications and Mobile Computing, pp. 797–802 (2006)
15. Wightman, P., Labrador, M.: A3: A topology control algorithm for wireless sensor networks. In: Proceedings of IEEE Globecom (2008)

16. Khan, M.A.M.: Distributed Approximation Algorithms for Minimum Spanning Trees and other Related Problems with Applications to Wireless Ad Hoc Networks. Ph.D. Dissertation, Purdue University (December 2007)
17. Saha, S.: Topology Control Protocols in Wireless Sensor Networks. MS Thesis, Texas A&M University-Kingsville (May 2014)

Elements of Difficulty Level in Mathematics

Taku Jiromaru, Tetsuo Kosaka, and Tokuro Matsuo

Abstract. It is important for each leader to know relationship of the reason why each learner mistakes the problem of Mathematics. But, it does not exist about previous research. Therefore, in this research, we collected answers of each learner for knowing element of difficulty level in Mathematics. And we identified 10 types. 10 types are "lack of understand(problem statement, Number & Symbol, Formula, Concept)", "circumstances of learner (inside)", "circumstances of learner (outside)", "Miscalculation", "Copy miss", "Lack of logical thinking (Deduction)" and "Lack of logical thinking (Induction)".

Keywords: Incorrect reason, arithmetic, mathematics, classification, difficulty.

1 Introduction

In some countries, a test has a profound influence on students' life. In Japan, high-school students who wish to go to high selection university have 3 ways to enter a university. It is the most major way to take an original test created by each university in addition to National Center Test for University Admissions, like Scholastic Achievement Test in United States or baccalaureat in France. For example, 84.7% of new students enter Yamagata University Faculty of Engineering used this way (Yamagata 2013).

The students can take the National Center Test for University Admissions once a year. And the students can take the original test only once a year. Can only once test evaluate each student's ability accurately? Shojima says "Current test is

Taku Jiromaru · Tetsuo Kosaka
Yamagata University, Faculty of Engineering, Yamagata, Japan
e-mail: jiro@om-edu.jp, tkosaka@yz.yamagata-u.ac.jp

Tokuro Matsuo
Advanced Institute of Industrial Technology, Tokyo, Japan
e-mail: matsuo@tokuro.net

© Springer International Publishing Switzerland 2015
R. Lee (ed.), *SNPD*,
Studies in Computational Intelligence 569, DOI: 10.1007/978-3-319-10389-1_7

overestimated. Current test has only resolution that divides about several stages."
(Shojima 2009).

The way of giving questions is one of the methods of measuring how learners
understand. Leaders are making and choosing for measuring learners understanding, but leaders make and choose questions by implicit knowledge because the
degree of difficulty for each question is not given by number.

Therefore, if it is possible to make the degree of difficulty of the problems in
mathematics, it will be used in a wide range of areas.

One of the ways of making a decision about the degree of difficulty for each
question is percentage of questions answered correctly (Maeda and Nishio 2000).
But, it has no mean to compare the degrees of difficulty between deferent categories of semantic structure (Riley and Greeno 1988). So, at least it has no mean to
decide the degrees of difficulty only percentage of questions answered correctly.

As another way of making a decision about the degree of difficulty for each
question, there is IRT (Item Response Theory) (Baker 1992). IRT is proposed as
the way to be able to account ability fairly and used TOEFL and TOEIC
(Tsukihara et al. 2008). But, in mathematics, there is less of a chance of used case
(Tsukihara et al. 2008).

Further, studies identified using the methods of the data mining difficulty is also, it is still under development presently believed to be necessary and a study on
the learning history data   (Baker 1992).

Therefore, in this research, we collected answers of each learner for knowing
element of difficulty. As well, "Difficulty" defines intractableness for solving each
question.

2 Methods of Making a Decision of Difficulty Level

For making a decision of difficulty level in mathematics, we were examined 3
types of method, "statistical approach", "subjective method", and "discovery of
the elements".

2.1 Statistical Approach

Statistical approach can find out whether the number is correct or not because we
can do additional test.

For example, Hirashima defined difficulty by percentage of questions answered
correctly (Hirashima et al. 1955), IRC defined difficulty by ability of each learner
and percentage of questions answered correctly(Riley and Greeno 1988), and Neural Test Theory(NTT), the method for standardizing test, defined difficulty by
self-organizing map (Kohonen 2000)(Shojima 2009).

But, in statistical approach, we can know conclusive difficulty level after examination designers make math problems, as exemplified by IRC.

So, this method cannot use when examination designer must hide these questions. For example, problems used admission test in Japanese university must be created originally and unreleased.

In addition, we have a data about IRC. Fig.1 and Fig.2 are the results of employment mock exam. In Japan, many companies do paper test as employment exam. Pieces of this exam are "Language area" and "Non-language area". "Language area" is including Japanese and English. "Non-language area" is including Mathematics and Science. Fig.1 is "Language area", and Fig.2 is "Non-language area". Vertical axis of Fig.1 and Fig.2 is ability of each learner, and Abscissa axis is point of each learner of employment mock exam. Ability of each learner calculated used the way of IRC.

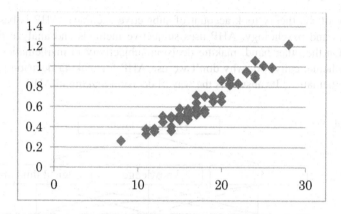

Fig. 1 Result of employment mock exam (Language area)

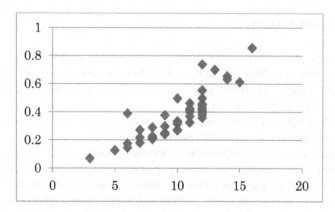

Fig. 2 Result of employment mock exam (Non-language area)

We calculated coefficient of correlation about Fig.1 and Fig.2. Coefficient of correlation about Fig.1 is 0.96075, and Fig.2 is 0.86258. These points are very high, but point of "Language area" is more than "Non-language area". This means we understand that this result shows that it is difficult for Mathematics to use IRC.

2.2   Subjective Methods

The method making decision subjectivity is commonly used. But, the value is not possible that there is credibility because there is no way to find out whether the value is correct or not. We have built educational system using this methods (Jiromaru 2012)

If a human decide difficulty level subjectivity, these stages may be reasonable. But, there are 2 problems. One is accuracy, another is resolution. These points have the potential of decreasing the resolution of test.

2.2.1   Accuracy of Difficulty Level

AHP is one of the theory took account of subjective accuracy. AHP is based on mathematics and psychology. AHP uses subjective methods and analyze mathematically. On the other hand, making decision subjectivity in mathematics don't analyze mathematically. So, why don't we use AHP for making decision subjectivity in mathematics? Its answer is the way of decision making in AHP.

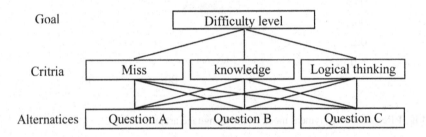

Fig. 3 A simple AHP hierarchy9

AHP (The Analytic Hierarchy Process) is a structured technique for dealing with complex decisions. Steps of AHP are as follow. At first, deciders break problem down into "goal", "criterions" and "alternatives". In particular, model of AHP represent hierarchy diagram like fig. 3.

Goal which is top layer is only one because of an end destination. Criterions set under goal, and alternatives set under criterions.

Next, decider calculates comparative importance between elements about each layer of criterions and alternatives. We use "pairwise comparisons" for collecting basic data. Pairwise comparisons aren't the method which determines to evaluate many elements all at once. It is the method which determines relative merits to take one pair methods.

Decider calculates proper vector and eigenvalue, and calculates Consistency Index (CI) using eigenvalue, and calculates importance of element using proper vector. If decider finishes calculating importance of all elements, decider can get final assessment of alternatives.

By using CI, AHP consider subjective methods and analyze mathematically. But, AHP has a problem. If criterions are change, result may be change (Belton and Gear 1985). So, it is important to set criterions mutually exclusive and collectively exhaustive.

2.2.2 Resolution of Difficulty Level

No one can divide an infinitum of stages. Shojima says "Current test is overestimated. Current test has only resolution that divides about several stages." (Shojima 2009). AHP, the method using subjective decision, also use several stages (Saaty 1994). In their papers, Shojima uses 10 stages, and AHP uses 9 stages (Shojima 2009) (Saaty 1994). There is no exist how many resolution human can know. But, perhaps the number is 9 or 10.

2.3 Discovery of the Elements

The research finding elements of difficulty level is many. For example, there is declination of learning (Shimizu and Ide 2003), or semantic structure of problem statement (Maeda and Nishio 2010), or computational complexity (Hirashima et al. 1955), and so on.

Raghubar et al. research about errors in multi-digit arithmetic and behavioral inattention in children with math difficulties (Raghubar et al. 2009).

But, there is not exist about a research about all elements of difficulty level.

2.4 Other Viewpoints

According research about difficulty level, there are other viewpoints.

For example, Soga et al. research about differences in physiological responses induced by mental tasks with different difficulty levels (Soga et al. 2009). Differences of difficulty level make a change about physiological responses and mental tasks, and it can calculate Skin Potential Level and Blood Pressure clearly if examinees don't feel anxiety.

3 The Reasons of Wrong Answer or Expression

3.1 How to Correct

We forecast that there is more than one reason. So, we corrected the reason of wrong answers or expression to 15 people. They are university students. The way of correcting the reasons shows Fig. 4.

One learner tries to solve problem, and one leader watch learner's action. If learner is wrong answer or expression, leader tells answer or expression is wrong. And leader asks learner why learner is wrong. At this time leader see not to make learner to feel psychological pressure.

If learner has no idea, leader advance an idea to learner. If learner has an idea, leader and learner entertain that the idea is right. If both leader and learner think that the idea is right, leader records the idea.

This way has possibilities not to be able to find idea. In this research, we don't have such case. But, if we have such case, more than one leader entertains the idea.

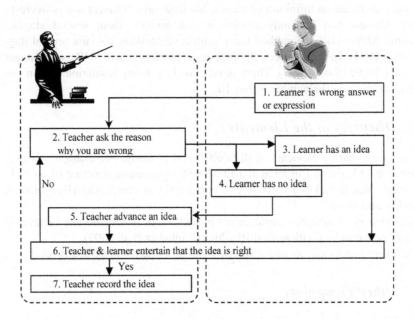

Fig. 4 Correcting reason of wrong answer or expression

3.2   *Arrangement of Data Collection*

To increase the accuracy of the collected data was performed prior arrangements as follows.

- In miscalculation, or do not understand the calculation method, I understand technique but able to tell exactly what made a mistake is difficult. Therefore, when miscalculation occurs, to point out that there is a miss by specifying a formula that leadership is a miscalculation to the learner, the learner calculated by without the help of a leader within 30 seconds it is counted as the only miscalculation if you notice the mistake.
- I cases are considered in No.6 in Figure 3, and repeat forever and No.4 or No.5 if leaders cannot present a draft of the incorrect reasons. Did not occur if such a survey in this paper, but if this happens, then consider leaders between pluralities why.

- If the physical situation is not normal or spiritual leader or learner, impossible to make accurate data is expected. In that case, to temporarily stop the data collection at the discretion of the leader. Once with fever, I paused once data collection in the deterioration of mental status in the collection of this data

4 Classification of Incorrect Reason

We have analyzed the papers and interviews with learners. We list quantity in Table 1.

Table 1 Analysis of questionnaires and interview

Student	Person	Incorrect reasons
Elementary school	4	342
Junior-high school	5	451
High school	6	432
College	5	442
University	15	368

4.1 Lack of Understanding

Yoshizawa says that misstep about Mathematics has 16 types (Yoshizawa 2006).
- Special treatment on 0 and 1
- The misunderstanding about the meaning of the symbols
- The realization that the shape of the representation may be different if it is the same as the mathematical
- How to use "and", "or" and "so"
- How to use "all" and "some"
- Negative X negative=positive
- A rough idea in the calculation
- Understanding of the meaning of the problem statement and description
- The order of action and movement
- Reverse of action and movement
- Abstraction to learn while lack of awareness of specific
- Examination of the validity of the expression deformation and application of official
- Error in the object to be compared with respect to the ratio
- Gap in understanding surface caused by the change of the unit and expansion of the subject to be handled
- Lack of graphical feeling
- Learning content needed Intuitive explanation

These types can show "lack of understanding".

We reclassify these 16 types through the position of incorrect reason in table 2.

Table 2 classification incorrect reason about lack of understanding

	lack of understanding	Items
1	problem statement	How to use "and", "or" and "so"
		How to use "all" and "some"
		Understanding of the meaning of the problem statement and description
2	Number & Symbol	Special treatment on 0 and 1
		The misunderstanding about the meaning of the symbols
3	Formula	The realization that the shape of the representation may be different if it is the same as the mathematical
		Negative X negative=positive
		The order of action and movement
		Reverse of action and movement
		Examination of the validity of the expression deformation and application of official
		Gap in understanding surface caused by the change of the unit and expansion of the subject to be handled
4	Concept	Examination of the validity of the expression deformation and application of official
		Error in the object to be compared with respect to the ratio
		Gap in understanding surface caused by the change of the unit and expansion of the subject to be handled
		Lack of graphical feeling
		Learning content needed Intuitive explanation

Therefore, lack of understanding has 4 types as with Table2.

4.2   Other Reasons

We summarize the case similar cases of incorrect reasons collected by this survey. Other reasons of "lack of understanding" occur 1156. These reasons have 3 types and 6 elements; we will show Table 3.

4.3   Elements of Difficulty Level

In Mathematics, all result can mark "correct" or "incorrect". Of course, it may be correct partly. But, no result can mark neither "correct" nor "incorrect". Therefore, we can call incorrect reason "Elements of difficulty level".

We formulate 10 elements as with Table 4 using Table 2 and 3.

Table 3 Classification of wrong reason

	Incorrect Reason	Example
1	Circumstances of learner	
	a) inner	Change in physiological phenomenon
		Physical sign
	b) outer	Change in circumstance
		Small calculation sheet
2	Miss	
	a) Miscalculation	Miss by mental calculation
	b) Copy miss	Handwriting is a mess
3	Lack of logical thinking	
	a) Deduction	Meaning shift
		Logic leap
		Order of logic confusion
	b) Induction	Duplication and leakage
		Generalized too early

Table 4 Elements of difficulty level

Areas	Elements
Lack of understanding	1. Problem statement
	2. Number & Symbol
	3. Formula
	4. Concept
Circumstances of learner	5. Inner
	6. Outer
Miss	7. Miscalculation
	8. Copy miss
Logical thinking	9. Deduction
	10. Induction

5 Discussion

5.1 Is Decision of Difficulty Level in Mathematics "Complex System"?

If "decision of difficulty level in Mathematics is complex system" is true, it is right way to use statistical approach. A complex system is commonly understood as any system consisting of a large number of interacting components (agents, processes, etc.) whose aggregate activity is non-linear (not derivable from the summations of the activity of individual components), and typically exhibits hierarchical self-organization under selective pressures. (Joslyn and Rocha 2000)

Complexity subtending system has 2 types. These are "disorganized complexity" and "Organized complexity" (weaver 1948). Characteristic of disorganized complexity is a number of variables, and characteristic of organized complexity is emergence.

In this research, we categorized elements of difficulty level in mathematics 10 types. So, this is not "disorganized complexity" if our research is true. On the other hand, we can't conclude about existence of emergence.

Robert et al. found out that "Microslips", fluctuation of moving for ungraspable consciously, is caused about one time a minute (Robert et al. 1990). If microslips have an impact difficulty level in mathematics, it can be complex system.

In this research, we cannot conclude whether complex system or not. In the future, we hope to research this theme.

5.2 Lack of Understanding

Hill et al. assembled various researches that effective teachers have unique knowledge of students' mathematical ideas and thinking, and they called assembled researches "Mathematical Knowledge for Teaching (MKT)" (Hill et al. 2008). Fig.5 shows MKT model. MKT model build for teachers and this research build for learners.

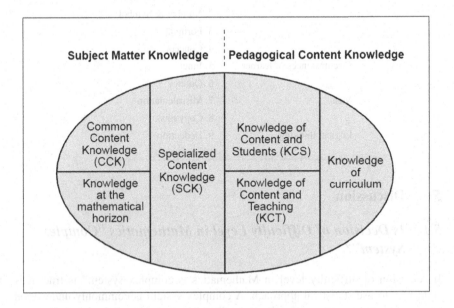

Fig. 5 Mathematical Knowledge for Teaching (Hill et al. 2008)

On the other hand, the research focused a particular of understanding in mathematics is very many. For example, Bardu and Beal said that English Learners' lower performance in math reflects the additional cognitive demands associated with text comprehension (Barb and Beal 2010).

5.3 Circumstances of Learner

About inner circumstance of learner, Meece et al. said that "Students' performance expectancies predicted subsequent math grades, whereas their value perceptions predicted course enrollment intentions." (Meese et al. 1990). This research confirms our research.

On the other hand, it is many researches with a central focus of Medical about outer circumstance of learner,

6 Conclusions

In this research, we correct incorrect reason for researching elements of difficulty level in Mathematics. We formulate 10 elements of difficulty level based on researching incorrect reason.

For making a decision of difficulty level in mathematics, we were examined 3 types of method, "statistical approach", "subjective method", and "discovery of the elements".

We decide to use "discovery of the elements". "statistical approach" need many data of each question. So, it does not use for new question. "subjective method" cannot show difficulty level strictly for decreasing resolution.

In this research, we find 4 types and 10 elements of difficulty level in mathematics. It makes a platform of difficulty level in Mathematics. There are many things about the research of each type or each element. It is necessary to refer to the huge quantity of previous research in a broad range of areas.

About difficulty level, Chalmars advocates "The problem that can build by algorism is easy" (Chalmars 1982), Searle advocates "there is no exist when algorism is performed the operation" (Searle 1982), and Penrose advocates "essence of consciousness has unreckonable process" (Penroose 1989). But Mogi advocates "These 3 opinion do not come at essential problem of human's cognition" (Mogi 2006) and we agree with Mogi's opinion. We are calculating data of incorrect reason, but simple and clear conclusions have not gotten yet. This theme is intimately-connected to nature of humanity. So, we will not be able to resolve easily, but we think that it is considered of value.

We forecast that conclusion will be using several methods like MKT, several methods is including this research, basic theories for each element, high cost-effectiveness collection methods of data, data analysis algorism, fuzzy system after reduced noise.

It is many topics of research and agendas about difficulty level in Mathematics. In the future, we will research about these topics.

References

1. Baker: Item Response Theory: Parameter Estimation Technique. Marcel Dekker, United States (1992)
2. Barbu, C.O., Beal, R.C.: Effects of Linguistic Complexity and Math Difficulty on Word Problem Solving by English Learners. International Journal of Education 2(2), 1–19 (2010)
3. Belton, V., Gear, T.: The legitimacy of rank reversal-A Comment. OMEGA the International Journal of Management Science 13(3), 14444 (1985)
4. Chalmers, D.J.: The conscious mind: In search of a fundamental theory. Oxford University Press, Oxford (1996)
5. Hirasihima, Y., Mori, T., Tani, T.: On the System of Numbers Considered from the Degree of Difficulty. Journal of JAPAN Society of Mathematical Education 37(10), 148–151 (1955)
6. Kohonen, T.: Self-Organizing Maps. Series in Information Sciences, pp. 1–521 (2000)
7. Maeda, M., Nishio, Y.: Research on Difficulty of Multiplication and Division Word Problems. Mathematics Education Research, 531–137 (2000)
8. Meece, J., Wigfield, A., Eccles, J.: Predictors of math anxiety and its influence on young adolescents' course enrollment intentions and performance in mathematics. Journal of Educational Psychology 82(1), 60–70 (1990)
9. Joslyn, C., Rocha, L.: Towards semiotic agent-based models of socio-technical organizations. In: Proc. AI, Simulation and Planning in High Autonomy Systems (AIS 2000), pp. 70–79 (2000)
10. Hill, H.C., Ball, D.L., Schilling, S.G.: Unpacking Pedagogical Content Knowledge: Conceptualizing and Measuring Teachers' Topic-Specific Knowledge of Students. Journal for Research in Mathematics Education 39(4), 372–400 (2008)
11. Jiromaru, T., Matsuo, T.: OMES: Employment support system for high education. CIEC, Computer & Education 32, 71–76 (2012)
12. Mogi, K.: Generation and qualia in brain. In: Suzuki, H. (ed.) Emergent and Conception of Intelligence, pp. 25–40. Ohmsha (2006)
13. Penrose, R.: Emperor's New Mind. Oxford University Press, Oxford (1989)
14. Raghubar, K., Cirino, P., Barnes, M., et al.: Journal of Learning Disabilities, 4356–4371 (2009)
15. Riley, N.S., Greeno, J.G.: Developmental Analysis of Understanding Language About Quantities and of Solving Problems. Cognition and Instruction 5(1), 49–101 (1988)
16. Saaty, L.S.: How to Make a Decision: The Analytic Hierarchy Process. Interfaces 24(6), 19–43 (1994)
17. Searle, J.: The Chinese room revisited. Behavioral and Brain Sciences 5(2), 345–348 (1982)
18. Shimizu, H., Ide, S.: Gaps in students' learning in elementary mathematics classes. Journal of Japan Society of Mathematical Education 85(10), 11–18 (2003)
19. Shojima, K.: Neural Test Theory: A Test Theory for Standardizing Qualifying Tests. The Journal of Institute of Electronics Information and Communication Engineers 92(12), 1014016 (2009)

20. Soga, C., Miyake, S., Wada, C.: Differences in Physiological Responses Induced by Mental Tasks with Different Difficulty Levels. The Japanese Journal of Ergonomics 45(1), 29–35 (2009)
21. Roberts, R.J., Varney, N., et al.: The neuropathology of everyday life: The frequency of partial seizure symptoms among normals. Neuropsychology 4(2), 65–85 (1990)
22. Tsukihara, Y., Suzuki, K., Hirose, H.: A small implementation case of the mathematics tests with the item response theory evaluation into an e-learning system. CIEC, Computer & Education 24, 70–76 (2008)
23. Weaver, W.: Science and complexity. American Scientist 36, 536–544 (1948)
24. Yamagata University Faculty of Engineering, Past entry-exam data (2013), http://www2.yz.yamagata-u.ac.jp/admission/admissiondata. html (accessed April 24, 2014)
25. Yoshizawa, M.: Classification of Students' Stumbles while Learning Mathematics. Journal of Japan Society of Mathematical Education 88(3), 228 (2006)

Identify a Specified Fish Species by the Co-occurrence Matrix and AdaBoost

Lifeng Zhang, Akira Yamawaki, and Seiichi Serikawa

Abstract. Today, the problem that invasive alien species threaten the local species is seriously happening on this planet. This happening will lost the biodiversity of the world. Therefore, in this paper, we propose an approach to identify and exterminate a specialized invasive alien fish species, the black bass. We combined the boosting method and statical texture analysis method for this destination. AdaBoost is used for fish detection, and the co-occurrence matrix is used for specified species identification. We catch the body texture pattern after finding the fish-like creature, and make a judgement based on several statistical evaluation parameter comes from co-occurrence matrix. Simulation result shows a reasonable possibility for identify a black bass from other fish species.

1 Introduction

Recently, a number of foreign creature was pointed as a specialized invasive alien species in Japan. This is because some inconsiderate pet owner discarded their breeding creature to the outdoor environment, or some fishing fun released the alien fish species into the river for fishing. Such a happening could destroy the ecosystem of native species, and is occurring world wide. This cause the creatures that not belong to the new place threaten the local creatures. When a stranger creature added into a balanced ecological circle, the condition of food sharing, speed of breeding would cause the native weak species disappeared fast if they are not be protected correctly. Currently, catching such an invasive alien species manually is main approach. It will spend enormous manpower and money once the invasive alien species founded their own life circle, and the effect is not expectable. To avoid

Lifeng Zhang · Akira Yamawaki · Seiichi Serikawa
Department of Electrical and Electronic Engineering, Kyushu Institute of Technology,
Fukuoka 804-8550, Japan
e-mail: {zhang,serikawa}@elcs.kyutech.ac.jp,
      yama@ecs.kyutech.ac.jp

© Springer International Publishing Switzerland 2015   99
R. Lee (ed.), *SNPD*,
Studies in Computational Intelligence 569, DOI: 10.1007/978-3-319-10389-1_8

such a happening, it is necessary to eradicate them before the invasive alien species explosively breed.

Once breeding is starting, to clean the black bass of all is difficult. Therefore, the efficient way to protect the ecological environment of native species is prevent the growth and disinfect it in a early stages if the invasive alien species has invaded. Proposed method intend to set up a camera in the water of ponds and lakes that do not have the invasive alien species such as black bass still. If the alien species invade, find them immediately. The approach that we think is first using AdaBoost to identify whether there is a fish exist in the picture of a camera. If there is a fish exist, then cut out the target area. Next, select several local area from the fish body, calculate the value of several parameters based on co-correlation matrix texture analysis. Then determine whether the target is a black bass or not.

In this study, we create a novel fish area detector for applying the AdaBoost. Then use the co-occurrence matrix for texture analysis to three kinds of fishes, black bass, carp and crucian, which are biological habitat same. From the evaluation value, we can finding the available parameters which can separate the target species from others by the body surface texture pattern.

In chapter 2, several target detection methods are introduced. In chapter 3, texture analysis methods are introduced, and the co-occurrence matrix method is selected for this work. Simulations for fish detection with AdaBoost and specified fish species identification are introduced in chapter 4 and finally chapter 5 is the conclusion.

2   Fish Area Detection

2.1   Overview of Target Detection

For the subject detection technique, template matching is a traditional way. In the template matching, a standard target pattern is manually predefined. Given an input image, the cross correlation value with the standard pattern is calculated for each local area separately. The existence of target is determined based on the correlation values. This method has the advantage of easy implementation. However, it has proven to be inadequate for creatures detection because it cannot effectively deal with the variation in scale, pose and shape. Furthermore, a lot of computing resource is need, the calculation speed is slow[1, 2, 3, 4].

Appearance-Based Methods is another target finding algorithm. Compare to the template matching methods, the template in appearance method are learned from the examples in the images but not predefined manually. Eigenfaces[5, 6, 7], Distribution-Based Methods[8, 9], Neural Networks[10, 11], Support Vector Machines[12, 13], Naive Bayes Classifier[14, 15, 16], Hidden Markov Model[17] are all developed for subject recognition or identification. But all of these approach are time consumption, not suit for a real time application until the algorithm below appeared.

In recent decades, an adaptive boosting algorithm called AdaBoost[18, 19] is invented. This boosting algorithm does not require any prior knowledge about the performance of the weak learning algorithm. In this method, many weak classifier consists a strong classifier with a learning process to find a suitable weightiness of each weak classifier. The detection speed is very fast and the performance is good for realtime application.

In this work, AdaBoost algorithm is adopt for fish area detection.

2.2 Mechine Learning Process

2.2.1 Sample Creation

AdaBoost is a excellent pattern detection algorithm, but in most case, it is used for face detection. There is no research use AdaBoost algorithm to detect a fish, thus we need to make a image database by ourself for study file creation.

In this work we collect 10 black bass images(Fig. 1) from internet or take a photo of experiment fish tank. And the processing tool is with OpenCV library.

Ordinarily, the study process needs more than 5000 positive images and 3000 negative images. It is impossible for us to collect such amount images, here we use the createsamples command included in OpenCV library to make 3000 formative sample images from these 10 images. Figure 1 show these image examples.

2.2.2 Training

In this work, machine learning process is also executed using an exist tool included in OpenCV library called haartraining. By modifying the option parameters of this command, finally, we got the study files in which recorded the weak classifier and the weighting coefficient. We do not introduce the Training process in detail because of the paper space. The reader can find the detail explanation on internet easily.

The summary of the steps are as following

a. Target Image Collection:
   In this work, only signal-object-taken images are selected, and the chosen image background is also clear. we then put such images in to one folder on the compute disk.
b. Negtive Image Selection:
   Several background images without the detection target include are collected. The size of the negative image larger than the positive image size so that each teaching image have a different background.
c. Postive Image Creation
   Applying the createsamples command with specify the target image and the negative images that you prepared in a. and b., the positive images are created automatically.
d. Postive Image Data Confirmation

Fig. 1 Fish examples for study

Before staring the Learning process, we check the created positive image by
human eyes and determine whether or not it is a reasonable data set for studying.
e. Processing of Machine Learning
    After getting a seasonal data set, we feed them in to a `haartraining` com-
    mand, finally we get the learned cascade feature file.
f. Object Detection
    Now we can detect the target object by using the feature file created by step e.

The detection result is shown in Fig. 2. All the fishes are detected correctly, and
the deferent species are not detected. Due to the experiment is executed under an
ideal environment, there are still thought many miss detection exist when for practical use. Here, we think this step can be correctly accomplished, then go to the next
steps.

3   Targeted Invasive Alien Fish Species Decision

In order to identify the extermination target species, black bass, from other fishes,
we proposed a quantifying the texture analysis approach from body surface pattern
of fish.

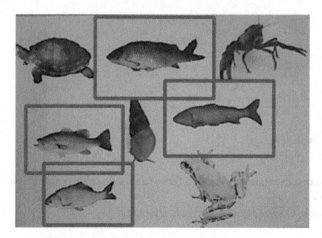

Fig. 2 Fish detection result with the learned feature file

3.1   Extraction of Texture Feature

Texture plays a very important role in visual information processing. It is thought that there are two texture level[20], structural texture and statistical texture[21, 22]. Considering from the standpoint of texture image analysis, more stronger the structure is, more easy for analysis. However , those nature texture is almost exists in statistical manner. And there is no way to analyze the texture in unique over the difference texture.

Broadly speaking the extraction of texture features of conventional can be devided in to:

(1) Extraction of statistical features.
(2) Analysis by local geometric feature.
(3) Analysis by fitting model..
(4) Structural analysis

Most of the analysis method is defined mathematically, which are not taken into account of correspondence between the visual psychology, but it is a quite effective way.

Texture features so strongly affected by the illumination of the image shooting, as a pretreatment, normalize the density of the image before extracting texture features is good in many cases.

As a typical calculation method of texture statistical characteristics, there are density histogram, co-occurrence matrix, the difference statistic, run length matrix, the power spectrum. Texture analysis is referred to as a N_{th} order statistical value relate to the combination of the concentration of point d at the certain position relationship.

Statistics introduced from the density histogram is 1st order statistics, statistics derived from co-occurrence matrix, difference statistics and power spectrum are the

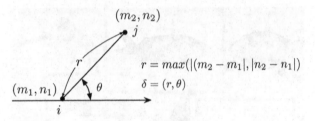

Fig. 3 Illustration of $P_\delta(i,j)$

2nd order statistics, statistics derived from run length matrix the amount is higher order statistics.

In this work, Co-occurrence matrix is adopted.

3.2 Co-occurrence Matrix

In Fig. 3Assuming $P_\delta(i,j)$ is the brightness probability of pixel $j = b(m_2, n_2)$ which has a displacement $\delta = (r, \theta)$ from pixel $i = b(m_1, n_1)$, the co-occurrence matrix of $(i, j = 1, 2, \cdots, n-1)$ can be derived. by calculating 11 types of feature value from co-occurrence matrix, the texture can be characterized by these values.

Let $P_x(i)$, $P_y(j)$, $P_{x+y}(i+j)$ and $P_{x-y}(|i-j|)$ defined as following,

$$P_x(i) = \sum_{j=0}^{n-1} P_\delta(i,j) \quad i = 0,1,...,n-1 \tag{1}$$

$$P_y(j) = \sum_{i=0}^{n-1} P_\delta(i,j) \quad j = 0,1,...,n-1 \tag{2}$$

$$P_{x+y}(i+j) = \sum_{i=0}^{n-}\sum_{j=0}^{n-1} P_\delta(i,j) \quad k = 0,1,...,2n-2 \tag{3}$$

$$P_{x-y}(|i-j|) = \sum_{i=0}^{n-}\sum_{j=0}^{n-1} P_\delta(i,j) \quad k = 0,1,...,n-1 \tag{4}$$

Angular second moment is given by

$$ASM = \sum_{i=0}^{n-1}\sum_{j=0}^{n-1} P_\delta(i,j)^2 \tag{5}$$

Contrast is given by

$$CNT = \sum_{k=0}^{n-1} k^2 \cdot P_{x-y}(k) \tag{6}$$

Correlation is given by

$$\text{CRR} = \frac{\sum_{i=0}^{n-}\sum_{j=0}^{n-1} i \cdot j \cdot P_\delta(i,j) - \mu_x \mu_y}{\sigma_x \sigma_y} \tag{7}$$

where

$$\mu_x = \sum_{i=0}^{n-1} i P_x(i)$$

$$\mu_y = \sum_{j=0}^{n-1} j P_x(j)$$

$$\sigma_x^2 = \sum_{i=0}^{n-1} (i - \mu_x)^2 P_x(i)$$

$$\sigma_y^2 = \sum_{j=0}^{n-1} (j - \mu_y)^2 P_y(j)$$

Sum of square variance is given by

$$\text{SSQ} = \sum_{i=0}^{n-1}\sum_{j=0}^{n-1} (i - \mu_x)^2 P_\delta(i,j) \tag{8}$$

Inverse difference moment is given by

$$\text{IDM} = \sum_{i=0}^{n-1}\sum_{j=0}^{n-1} \frac{1}{1 + (i-j)^2} \cdot P_\delta(i,j) \tag{9}$$

Sum average is given by

$$\text{SAV} = \sum_{k=0}^{2n-2} k \ldots P_{x+y}(k) \tag{10}$$

Sum variance is given by

$$\text{SVA} = \sum_{k=0}^{2n-2} (k - \text{SAV})^2 \ldots P_{x+y}(k) \tag{11}$$

Sum entropy is given by

$$\text{SEP} = -\sum_{k=0}^{2n-2} P_{x+y}(k) \cdot \log\{P_{x+y}(k)\} \tag{12}$$

Entropy is given by

$$\text{EPY} = \sum_{i=0}^{n-1}\sum_{j=0}^{n-1} \{P_\delta(i,j)\}^2 \tag{13}$$

Fig. 4 Body texture pattern selection area

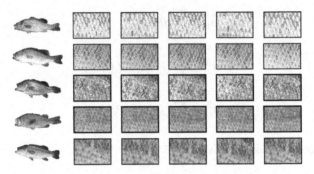

Fig. 5 Body texture pattern of black bass

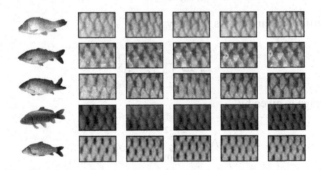

Fig. 6 Body texture pattern of crap

Difference variance is given by

$$\text{DVA} = \sum_{i=0}^{n-1} \left\{ k - \sum_{j=0}^{n-1} k \cdot P_{x-y}(k) \right\}^2 P_{x-y}(k) \tag{14}$$

Difference entropy id given by

$$\text{DEP} = \sum_{i=0}^{n-1} P_{x-y}(k) \cdot \log\{P_{x+y}(k)\} \tag{15}$$

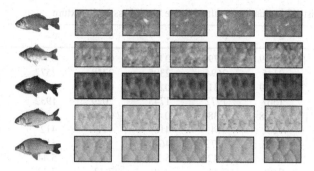

Fig. 7 Body texture pattern of crucian

Finally, there are things to note about the value of the feature be calculated. For example, the formula (5) ASM is used to measure whether a distribution which elements of the co-occurrence matrix is to position around. Thus, the uniformity of the texture can be measured. However, when an element of co-occurrence matrix are concentrated in the vicinity other than the diagonal, texture is not uniform.

As mentioned above, in some case it is unknown what you are physically measuring, it is necessary to pay attention to the value of the feature quantity calculated. Feature amount of the other is the same.

4 Simulations

By using the proposed method, an experiment was carried out to verify whether it is possible to identify the black bass from other fish species.

4.1 Fish Recognition Rate Experiment

In this experiment, using the Haar-Like as the image feature that is implemented in OpenCV, and used the AdaBoost as the learning algorithm, we produced an original fish detector. Then using the previously created fish detector to determine the error recognition rate and recognition rate. Because we have not a practical environment can be used, this experiment is performed under a ideal condition, The result is shown in Fig. 2. Although there was no error happened, it is necessary to enhance the study file to improve the availability in a practical environment.

4.2 Body Surface Texture Pattern Analysis

In this experiment, three kinds of freshwater fishes(black bass, carp, crucian) body texture are quantified by using the co-occurrence matrix, and then verify whether it is possible to identify the black bass from the two others. This time, using the lateral

Table 1 Statistical evaluation result of single black bass body texture

Evaluation	a	b	c	d	e	Average
ASM [$\times 10^{-3}$]	0.819	0.790	0.824	0.756	0.786	0.795
CNT	2145	2694	2395	2748	2572	2511
CRR	0.319	0.315	0.317	0.325	0.331	0.322
SSQ	1567	1972	1743	2029	1932	1849
IDM [$\times 10^{-1}$]	0.448	0.380	0.422	0.448	0.413	0.422
SAV	422	414	418	411	412	415
SVA	4158	5172	4616	5395	5122	4893
SEP	5.34	5.42	5.38	5.45	5.44	5.41
EPY	7.39	7.41	7.39	7.43	7.42	7.4
DVA	927	1140	1053	1197	1119	1087
DEP	4.51	4.63	4.56	4.62	4.59	4.58

Table 2 Statistical evaluation result of single Carp body texture

Evaluation	a	b	c	d	e	Average
ASM [$\times 10^{-3}$]	0.657	0.611	0.623	0.614	0.596	0.620
CNT	206	326	290	244	326	278
CRR	836	989	953	1102	956	967
SSQ	1283	1373	1300	1226	1317	1300
IDM[$\times 10^{-1}$]	0.718	0.600	0.631	0.617	0.594	0.632
SAV	336	290	302	315	307	310
SVA	3146	3646	3522	4203	3501	3604
SEP	5.34	5.40	5.39	5.47	5.39	5.40
EPY	7.43	7.48	7.47	7.48	7.50	7.47
DVA	66	105	93	78	100	88
DEP	3.39	3.61	3.56	3.47	3.60	3.52

peripheral image of the chest fin as shown by the red frame in Fig. 4. for each kind of fish, the body texture pattern shown in Figures 4-6 are used. From the side around the chest fin of each image, five pieces of the body texture image of 60×40 pixel was cut from each image.

Table $1 \sim 3$ show the quantified value of co-occurrence matrix for one single black bass, crap, crucian body texture on different body positions respectively. The last column is average value of different body position. The value in each column is relatively stable. this means such parameters can be used for texture evaluation.

Table $4 \sim 6$ show the quantified average value of co-occurrence matrix for each five black bass, crap, crucian body texture on different body positions respectively. Some of the parameters show the obviously difference for three kinds of fishes.

Table 7 shows the quantified feature value range of black bass, crap, crucian respectively. And we can find a available feature value to identify the black bass from two others.

Table 3 Statistical evaluation result of single crucian body texture

Evaluation	a	b	c	d	e	Average
ASM[$\times 10^{-2}$]	0.123	0.117	0.170	0.127	0.170	0.141
CNT	85	77	67	95	61	77
CRR	0.773	0.902	0.816	0.856	0.833	0.836
SSQ	186	401	183	335	184	258
IDM	0.1528	0.169	0.206	0.174	0.209	0.182
SAV	263	243	228	249	225	242
SVA	666	1515	667	1240	674	953
SEP	4.63	4.86	4.53	4.74	4.53	4.66
EPY	6.89	6.95	6.61	6.91	6.61	6.79
DVA	39	38	39	51	35	40
DEP	2.93	2.86	2.69	2.92	2.67	2.81

Table 4 Average statistical evaluation result of five black basses body texture

Evaluation	a	b	c	d	e
ASM[$\times 10^{-3}$]	0.795	0.489	0.489	0.542	0.515
CNT	2511	2626	2144	1129	1213
CRR	0.322	0.254	0.407	0.326	0.534
SSQ	1849	1755	1816	840	1300
IDM[$\times 10^{-1}$]	0.422	0.247	0.286	0.396	0.368
SAV	415	339	320	291	297
SVA	4893	4394	5121	2225	3993
SEP	5.41	5.50	5.58	5.20	5.46
EPY	7.41	7.65	7.65	7.58	7.61
DVA	1087	973	800	430	482
DEP	4.58	4.62	4.53	4.22	4.24

Table 5 Average statistical evaluation result of five craps body texture

Evaluation	a	b	c	d	e
ASM[$\times 10^{-3}$]	0.620	0.552	0.569	0.944	0.588
CNT	278	375	371	88	551
CRR	0.856	0.893	0.871	0.866	0.835
SSQ	967	1765	1447	343	1688
IDM[$\times 10^{-1}$]	0.632	0.711	0.688	1.09	0.659
SAV	310	245	237	102	233
SVA	3604	6659	5430	1281	6163
SEP	5.40	5.69	5.57	4.93	5.56
EPY	7.47	7.56	7.54	7.12	7.52
DVA	88	157	142	30	222
DEP	3.52	3.66	3.68	2.98	3.85

Table 6 Average statistical evaluation result of five crucians body texture

Evaluation	a	b	c	d	e
ASM[×10⁻²]	0.141	0.084	0.108	0.118	0.173
CNT	77	103	62	76	47
CRR	0.836	0.900	0.904	0.803	0.862
SSQ	258	521	326	196	174
IDM	0.182	0.128	0.152	0.144	0.191
SAV	242	251	146	356	310
SVA	953	1984	1227	706	649
SEP	4.66	5.13	4.91	4.64	4.55
EPY	6.79	7.23	7.01	6.93	6.62
DVA	40	41	24	31	20
DEP	2.81	3.05	2.81	2.91	2.65

Table 7 Average statistical evaluation result of three kinds of fishes

Evaluation	Black bass	Crap	Crucian
ASM[×10⁻³]	0.8~0.5	0.9~0.6	1.7~0.80
CNT	2626~1129	551~88	103~47
CRR	0.534~0.254	0.892~0.835	0.903~0.802
SSQ	1849~840	1765~343	521~174
IDM[×10⁻¹]	0.42~0.24	1.08~0.63	1.90~1.28
SAV	415~291	310~102	1984~649
SVA	5121~2225	6659~1281	1984~649
SEP	5.58~5.20	5.68~4.93	5.13~4.54
EPY	7.65~7.40	7.56~7.12	7.22~6.62
DVA	1087~430	222~30	41~20
DEP	4.61~4.21	3.84~2.97	3.05~2.65

From Table 7, the CNT value of black bass is more than 1000 which is larger than crap and crucian. CRR value is smaller than 0.6 which is little than the others. Furthermore, the DEV, DEP value of black bass is larger than crap and crucian, and IDM is small. We also can combine some of the parameters for a more stable identification. Or if we only need to separate the black bass from crucian, the SSQ, SVA, SEP and EPY also can be used.

5 Conclusion

In this study, we propose a specific method of fish species by co-occurrence matrix and AdaBoost, it was verified in a ideal condition. The fish area detection learning file for AdaBoost is made originally, and it is work well. For identify the target species from others, the feature amount of fish body texture is calculated from co-occurrence matrix. And by setting a threshold value, it is possible to identify the specialized invasive alien fish species from other local fishes. In future, we need to verify this work under a practical environment. And implement this method to a under water robots for under water surveillance.

References

1. Darrell, T., Gordon, G., Harville, M., Woodfill, J.: Integrated Person Traching Using Stereo, Color, and Pattern Detection. Int'l J. Computer Vision 37(2), 175–185 (2000)
2. Govindaraju, V.: Locating Human Faces in Photographs. Int'l J. Computer Vision 19(2), 129–146 (1996)
3. Govindaraju, V., Sher, D.B., Srihari, R.K., Srihari, S.N.: Locating Human Faces in Newspaper Photographs. In: Proc. IEEE Conf. Computer Vision and Pattern Recognition, pp. 549–554 (1989)
4. Govindaraju, V., Srihari, S.N., Sher, D.B.: A Computational Model for Face Location. In: Proc. Third IEEE Int'l Conf. Computer Vision, pp. 718–721 (1990)
5. Rahman, A., Rahman, M.N.A., Safar, S., Kamruddin, N.: Human Face Recognition: An Eigenfaces Approach. In: International Conference on Advances in Intelligent Systems in Bioinformatics. Atlantis Press (2013)
6. Shakhnarovich, G., Moghaddam, B.: Face recognition in subspaces. In: Handbook of Face Recognition, pp. 19–49. Springer, London (2011)
7. Kothari, A., Bandagar, S.M.: Performance and evaluation of face recognition algorithms. World Journal of Science and Technology 1(12) (2012)
8. Roth, M.: Peter and W. Martin, Survey of appearance-based methods for object recognition. Inst. for Computer Graphics and Vision, Graz University of Technology, Austria. Technical Report ICGTR0108, ICG-TR-01/08 (2008)
9. Kim, T.K., Jose, K., Roberto, C.: Discriminative learning and recognition of image set classes using canonical correlations. IEEE Transactions on Pattern Analysis and Machine Intelligence 29(6), 1005–1018 (2007)
10. Krizhevsky, A., Ilya, S., Geoffrey, E.H.: Image Net Classification with Deep Convolutional Neural Networks. In: NIPS, vol. 1(2) (2012)
11. Er, M.J., Chen, W., Wu, S.: High-speed face recognition based on discrete cosine transform and RBF neural networks. IEEE Transactions on Neural Networks 16(3), 679–691 (2005)
12. Dalal, N., Triggs, B.: Histograms of oriented gradients for human detection. In: IEEE Computer Society Conference on Computer Vision and Pattern Recognition, CVPR 2005, vol. 1, pp. 886–893 (2005)
13. Maldonado-Bascon, S., Lafuente-Arroyo, S., Gil-Jimenez, P., Gomez-Moreno, H., Lopez-Ferreras, F.: Road-sign detection and recognition based on support vector machines. IEEE Transactions on Intelligent Transportation Systems 8(2), 264–278 (2007)
14. Boiman, O., Shechtman, E., Irani, M.: In defense of nearest-neighbor based image classification. In: IEEE Conference on Computer Vision and Pattern Recognition, CVPR 2008, pp. 1–8 (2008)
15. Bay, H., Ess, A., Tuytelaars, T., Van Gool, L.: Speeded-up robust features (SURF). Computer Vision and Image Understanding 110(3), 346–359 (2008)
16. Boiman, O., Shechtman, E., Irani, M.: In defense of nearest-neighbor based image classification. In: Computer Vision and Pattern Recognition, CVPR 2008, pp. 1–8 (2008)

17. Nguyen, N.T., Hung, P.D.Q., Venkatesh, S., Bui, H.: Learning and detecting activities from movement trajectories using the hierarchical hidden Markov model. In: IEEE Computer Society Conference on Computer Vision and Pattern Recognition, CVPR 2005, vol. 2, pp. 955–960 (2005)
18. Serre, T., Wolf, L., Poggio, T.: Object recognition with features inspired by visual cortex. In: IEEE Computer Society Conference on Computer Vision and Pattern Recognition, CVPR 2005, vol. 2, pp. 994–1000 (2005)
19. Laptev, I.: Improvements of Object Detection Using Boosted Histograms. In: BMVC, vol. 6, pp. 949–958 (2006)
20. Zujovic, J., Pappas, T.N., Neuhoff, D.L.: Structural similarity metrics for texture analysis and retrieval. In: IEEE International Conference on Image Processing (ICIP), pp. 2225–2228 (2009)
21. Jafari-Khouzani, K., Soltanian-Zadeh, H.: Rotation-invariant multiresolution texture analysis using Radon and wavelet transforms. IEEE Transactions on Image Processing 14(6), 783–795 (2005)
22. Liu, G.H., Yang, J.Y.: Image retrieval based on the texton co-occurrence matrix. Pattern Recognition 41(12), 3521–3527 (2008)

Constraint-Based Verification of Compositions in Safety-Critical Component-Based Systems

Nermin Kajtazovic, Christopher Preschern, Andrea Höller, and Christian Kreiner

Abstract. Component-based Software Engineering (CBSE) is currently a key paradigm used for building safety-critical systems. Because these systems have to undergo a rigorous development and qualification process, one of the main challenges of introducing CBSE in this area is to ensure the integrity of the overall system after building it from reusable components. Many (formal) approaches for verification of compositions have been proposed, and they generally focus on behavioural integrity of components and their data semantics. An important aspect of this verification is a trade-off between scalability and completeness.

In this paper, we present a novel approach for verification of compositions for safety-critical systems, based on data semantics of components. We describe the composition and underlying safety-related properties of components as a Constraint Satisfaction Problem (CSP) and perform the verification by solving that problem. We show that CSP can be successfully applied for verification of compositions for many types of properties. In our experimental setup we also show how the proposed verification scales with regard to size of different system configurations.

Keywords: component-based systems, safety-critical systems, compositional verification, constraint programming.

1 Introduction

Safety-critical systems drive the technical processes in which failures can cause catastrophic consequences for humans and the operating environment. Automotive, railway and avionics are exemplary domains here, just to name few. In order to

Nermin Kajtazovic · Christopher Preschern · Andrea Höller · Christian Kreiner
Institute for Technical Informatics, Graz University of Technology, Austria
e-mail: {nermin.kajtazovic,christopher.preschern,andrea.hoeller,
        christian.kreiner}@tugraz.at

© Springer International Publishing Switzerland 2015
R. Lee (ed.), *SNPD*,
Studies in Computational Intelligence 569, DOI: 10.1007/978-3-319-10389-1_9

make these systems acceptably safe, their hardware/software engineering has to be rigorous and quality-assured.

Currently, rapid and continuous increase of system's complexity is one of the major challenges when engineering safety-critical systems. For instance, the avionics domain has seen an exponential growth of software-implemented functions in the last two decades (Butz (-)), and a similar development has also occurred in other domains with a focus on mass production (Kindel and Friedrich (2009)). In response, many domains have shifted towards using component-based paradigm (Crnkovic (2002)). The standards such as the automotive AUTOSAR and IEC 61131/61499 for industrial automation are examples of widely used component systems. This paradigm shift enabled the improvement in reuse and reduction of costs in development cycles. In some fields, the modularity of the system structure is utilized to distribute the development across different roles, in order to perform many engineering tasks in parallel (e.g. the automotive manufacturers are supplied by individually developed middleware and devices which can run their applications).

However, the new paradigm also introduced some new issues. One of the major challenges when applying CBSE is to ensure the integrity of the system after building it from reusable parts (components). The source of the problem is that components are often developed in the isolation, and the context in which they shall function is usually not considered in detail. In response, it is very difficult to localize potential faults when components are wired to form a composition – an integrated system (Gössler and Sifakis (2005)), even when using quality-assured components. The focus of the current research with regard to this problem is to enrich components with properties that characterize their correct behavior for particular context, and in this way to provide a basis for the design-time analysis or verification[1] of compositions (Clara Benac Earle et al (2013)).

This verification is also the subject of consideration in some current safety standards. For instance, the ISO 26262 standard defines the concept Safety Element out of Context (SEooC), which describes a hardware/software component with necessary information for reuse and integration into an existing system. Similarly, the Reusable Software Components concept has been developed for systems that have to follow the DO-178B standard for avionic software. These concepts both share the same kind of strategy for compositional verification: contract-based design. Each component expresses the assumptions under which it can guarantee to behave correctly. However, the definition of the specific contracts, component properties and validity criteria for the composition is left to the domain experts.

From the viewpoint of the concrete and automated approaches for compositional verification and reasoning, many investigations have focused on behavioural integrity, i.e. they model the behaviour of the components and verify whether the composed behaviours are correctly synchronized (de Alfaro and Henzinger (2001)), (Basu et al (2011)). On the other side, compositions are often made based on data semantics shared between components (Benveniste et al (2012)). Here, the correct behaviour is characterized by describing valid data profiles on component interfaces.

---

[1] In the remainder of this paper, we use the term *verification* for static, design-time verification (cf. static analysis (Tran (1999))).

In both cases, many properties can be required to describe a single component and therefore scalability of the verification method is crucial here.

In this paper, we present a novel approach for verification of compositions based on the data semantics shared between components. We transform the modelled composition along with properties into a Constraint Satisfaction Problem (CSP), and perform the verification by solving that problem. To realize this, we provide the following contributions:

- We define a component-based system that allows modelling properties within a complete system hierarchy.
- We define a structural representation of our modelled component-based system as a CSP, which provides us a basis to verify the preservation of properties.
- We realize the process that conducts the transformation of the modelled component-based system into a CSP and its verification automatically.

The CSP is a way to define the decision and optimization problems in the context of Constraint Programming paradigm (CP) (Apt (2003)). Using this paradigm for our component-based system, many types of properties can be supported. Also, various parameters that influence the scalability of the verification can be controlled (used policy to search for solutions for example). In the end of paper, we discuss the feasibility of the approach with regard to its performance.

The remainder of this paper is organized as follows: Section 2 describes the problem statement more in detail and gives some important requirements with regard to modelling a system. In Section 3 the proposed verification method is described. Section 4 describes the experimental results. A brief overview of relevant related work is given in Section 5. Finally, concluding remarks are given in Section 6 .

2 Problem Statement

Properties are an important means to characterize functional and extra-functional aspects of components. Safety, timing and resource budgets are examples here, just to name few (Sentilles et al (2009)). Recently, they get more and more attention in the safety community, since efficient (an practical) reuse methods are crucial in order to reduce costs in development cycles and costs for certification of today's safety-critical systems (i.e. their extensive qualification process). In this section, we give an insight into the main challenges when using properties to verify compositions, and based on these challenges, we outline the main objectives that we handle in this paper.

2.1 Motivating Example

In our work, we address properties that in general describe data semantics. To clarify this, let us consider now the example from Figure 1. The system in this figure shows the composition of four components that form the automotive engine control application on a higher abstraction level. The basic function of this application is

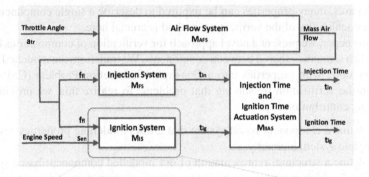

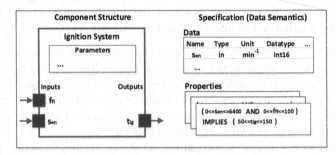

Fig. 1 Motivating example: a component-based system of automotive engine control function, adopted from (Frey (2010)) (top), and detailed view of the component Ignition System (structure and specification, bottom)

to decide when to activate the tasks of the fuel injection and ignition (Frey (2010)). To do this, the application takes the sensed values of the air flow volume, current speed and some parameters computed from the driver's pedal position. In a typical automotive development process[2], the system structure from figure is made based on stepwise decomposition of top-level requirements, having several intermediate steps such as the functional and technical system architecture with several levels in the hierarchy. Let us assume now that involved components are already developed, eventually for the complete car product line, and are stored in some repository. Let us further assume that we have a top-level requirement with regard to the engine timing for particular car type, which states the following:

The minimal allowed time delay between the task of the fuel injection and ignition shall be greater than 40 ms.

The main contributors to this requirement are software components M_{AFS}, M_{FS}, M_{IS}, M_{IIAS}, and their execution platform (e.g. concrete mapping of components on real-time tasks, task configurations, and other). In order to satisfy this timing property, the developer has to analyze the specification for each component in order to find the influence of the component behaviour on that property. The example of such

---
[2] Note that we do not limit our approach to automotive domain.

a specification is given in Figure 1, bottom. Here, the context for the component Ignition System is defined in terms of the syntax and semantics related to component inputs, outputs and parameters. With the properties shown below, the concrete behavior can be roughly described – in this example, for certain intervals of inputs, the component can guarantee that the output t_{ig} lies within the interval $[50, 150]$ (note that pseudo syntax is used here). When building compositions based on such properties, the developer has to consider their influence on the remaining, dependent components – in this case, it has to be decided whether the M_{IIAS} component can accept such values of the t_{ig} and what should components M_{FS} and M_{AFS} provide so that higher delay than $40ms$ between t_{ig} and t_{in} can be achieved. This can be very tedious and error prone task when doing it manually, because of the following reasons:

- Many components may be required to build a complete system, depending on their granularity. For example, current automotive systems comprise several hundreds of components, and many of them may depend on each other (Kindel and Friedrich (2009)).
- Some components that directly influence the safety-critical process are usually certified, i.e. developed according to rigorous rules from safety standards. Because of costs for such a certification, the practice is to develop components for different context and to certify them just once (e.g. to support different engine types in our example). In response, many properties have to be defined for a single component to capture all context information.

The main problem here is how to define and to inter-relate all properties thorough the complete system hierarchy in a way that the preservation of properties of all components can be verified automatically? Another problem is how to complete with such a verification in a "reasonable time"?

2.2 Modelling and Verification Aspects

To narrow the problem statement above, very important prerequisite to structure properties within a system hierarchy consistently is to define basic relations among them. For example, properties of the component M_{IS} are related with properties of the component M_{IIAS}, because of direct connections between their output and input variables. On the other hand, properties of all four components influence the semantics of the mentioned top-level property. We summarize different types of these relations as following:

- *Composition*: hierarchical building of composed properties based on their contained properties (e.g. the top-level timing property is composed of properties contained in components M_{AFS}, M_{IS}, M_{FS} and M_{IIAS}). We discuss this later in more detail.
- *Refinement/abstraction*: properties characterize the component behaviour at certain abstraction level. With refined properties, more specialized behaviours can

be described. For example, the property in Figure 1 may include some additional parameters to define conditions for the t_{ig} more precisely.

- *Alternatives*: properties may have alternative representations for different context (e.g. the Injection System component M_{IS} can provide different properties for different engine types).

These relations have to be supported when modelling a component-based system and they have to be considered when such a system has to be verified.

3 Constraint-Based Verification

In this section, we describe the proposed approach for compositional verification. To get a rough image of our approach, we summarized the basic steps in Figure 2

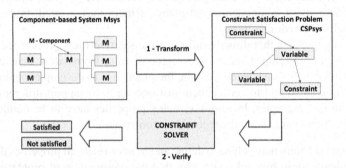

Fig. 2 Overview of the proposed verification method: (1) transformation of the component-based system M_{sys} into the CSP representation CSP_{sys}, (2) verification of the composition CSP_{sys} by solving a CSP

that we perform to conducts the verification process. The input to the verification is a modelled component-based system, enriched with properties – M_{sys} in figure. This model is further transformed into a Constraint Satisfaction Problem (CSP) – CSP_{sys} in figure, which is a network of inter-connected variables and constraints (we discuss this later). The CSP model is processed by the constraint solver, i.e. a tool to solve the CSPs, in order determine the preservation of all properties in the system. As a result, we get a decision about such a preservation. In addition, we get concrete values of data (i.e. inputs, outputs, parameters), for which properties are preserved. All steps in the process are performed automatically.

In the following, we describe how we defined each model described above. We first give some basic assumptions for our system M_{sys}. Then we describe the main elements of that system, including properties. In the end, we describe its representation as a CSP.

3.1 General: Components and Compositions

In our system, we define a component M as follows:

$$M := \langle \Sigma^{in}, \Sigma^{out}, \Sigma^{par}, M_c \rangle \tag{1}$$

, where Σ^{in}, Σ^{out}, and Σ^{par} are inputs, outputs and parameters respectively (i.e. Σ-alphabets define input, output and parameter variables in terms of datatypes, values, and some additional attributes), whereby M_c is an optional set of contained components, and is defined according to relation (1). To clarify this, we distinguish between following two types of components:

- *Atomic components*: components that can not be further divided to form hierarchies, i.e. components for which $M_c = \emptyset$. They perform the concrete computation. The Ignition System for example may contain many atomic components, such as integrators, limiters, simple logical elements and other.
- *Composite components*: hierarchical components that may contain one or more atomic and composite components, i.e. $M_c \neq \emptyset$. Note that we use the term *composition* to indicate composite components, which also may represent a complete component-based system (cf. our system in Figure 1).

The component model introduced above is typical for data-flow systems such as the ones modelled in the Matlab Simulink for example. Similar models are used when considering properties for resource budgets (Benveniste et al (2012)).

3.2 Modelling Compositions Enriched with Properties

As illustrated in Figure 1, properties are defined as expressions over component variables. In order to be able to interpret these expressions during the verification, we formulate them in a SMT form[3]: each expression can be represented in terms of basic symbols, such as $0, 1, ...s_{en}, ..., +, -, /, ...min$. Using this form, various expressions can be supported for our system, including logical, arithmetic, and other. The property from Figure 1 for instance, $(0 \geq s_{en} \leq 6400) \wedge (0 \geq f_{fl} \leq 100)$, conforms to the SMT form.

In order to link properties throughout the system hierarchy with regard to three basic relations introduced in Section 2.2, we encapsulate them in assume/guarantee (A/G) contracts. According to the general contract theory in (Benveniste et al (2012)), a contract C is a tuple of assumption/guarantee pairs, i.e.:

$$C := \langle \Sigma, A, G \rangle, \tag{2}$$

where A and G are expressions over sets of variables Σ. In this way, we can split properties for each component in (a) part that has to be satisfied, i.e. *assumptions*, and (b) part that is guaranteed if assumptions hold, i.e. *guarantees*. For example, the

---

[3] Syntax in SMT (Satisfiability Modulo Theories) allows to define advanced expressions, e.g. on integers, reals, etc.

top-level contract C_{II} for our system in Figure 1 guarantees the $40ms$ delay under assumptions that the rotational speed s_{en} and values for the throttle angle a_{tr} are within certain ranges:

$$C_{II} = \begin{cases} \text{variables} \begin{cases} \text{inputs} & s_{en}, a_{tr} \\ \text{parameters} & - \\ \text{outputs} & t_{in}, t_{ig} \end{cases} \\ \text{types} \quad s_{en}, a_{tr}, t_{in}, t_{ig} \in \mathbb{N} \\ \text{assumptions} \ (0 \geq s_{en} \leq 6400) \wedge (0 \geq a_{tr} \leq 100) \\ \text{guarantees} \quad t_{ig} - t_{in} > 40 \end{cases}$$

Based on this structure, we can link properties between dependent components in a similar way it is done when wiring components using connectors (i.e. links between their input/output variables). Figure 3 shows our example system modelled using contracts. Every component provides certain guarantees which stay in relation to assumptions of dependent components. These components in turn provide guarantees based on their own assumptions, and so forth. In this way, all properties within a system hierarchy can be linked together. In Figure 3, we have also highlighted different types of relations between contracts, required to build such a hierarchy (see Section 2.2). These are:

- *Composition*: two contracts can interact when after connecting their guarantees and assumptions both contracts can function correctly (we discuss this in more detail in Section 3.3). We use the operator $\otimes$ to define a composition (Benveniste et al (2012)). An example of such relations is shown in Figure 3, where contracts C_{FS}, C_{IS}, and C_{IIAS} form a composite contract, i.e. $(C_{FS} \otimes C_{IS}) \otimes C_{IIAS}$.
- *Refinement/abstraction*: similar to refinement of properties, contracts refine other contracts in terms of refined assumptions and guarantees. We use the operator $\preceq$ for this relation. The top-level contract C_{II} has such a relation with the contained

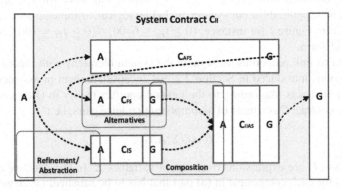

Fig. 3 The Engine Controller system represented using contracts and their basic relations (A – assumptions, G – guarantees, C – contracts)

contracts, i.e. $(C_{FS} \otimes C_{IS}) \otimes C_{IIAS}) \preceq C_{II}$. Note that only the relation with the contract C_{IS} is highlighted here.

- *Alternatives*: when designing components for more than one context, each new context is described in a separated contract. Contracts that describe the same property for different context are alternatives. In example in Figure 3, any of contained contracts may have alternatives – here, we just highlighted C_{FS} to indicate that it may have alternative contracts.

Based on definitions for contracts and their relations, we can now define the top-level system/composition contract, C_{sys}, as follows:

$$C_{sys} := (\otimes_{i \in \mathbb{N}} C_i) \tag{3}$$

, i.e. a hierarchical composition of contracts C_i, where C_i represents further composition according to relation (3).

Finally, to relate contracts with components, i.e. the concrete implementations of contracts, we extend the relation (1) as follows:

$$M := \langle \Sigma^{in}, \Sigma^{out}, \Sigma^{par}, C_c, M_c \rangle \tag{4}$$

, where C_c is a set of contracts that the component M can implement. Based on this relation, any implementation of the C_{sys} contract represents a complete component-based system or a top-level composition. We identify this implementation as M_{sys} and use it later as a basis to define our CSP.

3.3 Ensuring Correctness of Compositions

For our component-based system defined previously, two contracts C_1 and C_2 can form a composition (i.e. can be integrated) when their connected assumptions/guarantees match in the syntax of their variables (i.e. datatypes, units, etc.), and when following holds:

$$G(C_1) \subseteq A(C_2) \tag{5}$$

In other words, the contract C_1 shall not provide values not assumed by the contract C_2. This relation is a basis in our CSP to verify the complete composition.

3.4 Composition as a Constraint Satisfaction Problem

Now, we describe how we define the composition M_{sys} as a CSP. We name our CSP representation of M_{sys} as CSP_{sys}, and define it as follows:

$$CSP_{sys} := \langle X_{CSP}, D_{CSP}, C_{CSP} \rangle \tag{6}$$

, where X_{CSP} is a finite set of variables, D_{CSP} their domains (datatypes, values), and C_{CSP} a set of constraints related to variables and constraints in C_{CSP}. In other words, the CSP represents a network of variables inter-connected with each other using constraints. The constraints set variables in relations using some operators, and in this way they form expressions. Various types of expressions can be used to define constraints (e.g. Boolean, SMT – depending on supported features of the solver). The solution of the CSP_{sys} is a set of values of X_{CSP} for which all constraints C_{CSP} are satisfied. The constraint solver performs the task of finding solutions.

In order to represent the composition M_{sys} in a CSP, we need to map the top-level contract structure ((sub-)contracts, variables, and A/G expressions) into the CSP constructs mentioned above. Important aspects of this representation are CSP definitions for (1) a type system, (2) A/G expressions or properties, (3) the structure of components and contracts and (4) the structure of compositions. We can now turn to these representations.

3.4.1 Type System

The CSP tools, i.e. constraint solvers, usually provide the support for several domains to represent various types of variables. Integers, reals, and sets are examples here, just to name few. In order to avoid type castings between modelled system M_{sys} and CSP_{sys}, we use the same domains for both M_{sys} and CSP_{sys}. Another reason is that the time needed for the constraint solver to solve the CSP strongly depends on a particular domain. For example, there is a significant difference in runtime when dealing with real numbers instead of integers. Therefore, we use integers for both system representations, i.e. M_{sys} and CSP_{sys}.

3.4.2 A/G Expressions (Properties)

Concerning the representation of values of variables in the CSP, limits have to be set on their intervals. The intervals are possible search space for the solver, and can have significant influence on solver's runtime. It is therefore important to limit the variables on smallest possible intervals.

In our CSP_{sys}, each variable which is used in an expression is represented by two CSP variables: one indicating the begin of the interval, another one for the end of that interval. The size of this interval is determined based on intervals defined in expressions. For example, the variable s_{en} in the expression $(0 \geq s_{en} \leq 6400)$ is limited on the interval $[0, 6400]$. The reason for using two CSP variables here is that solving the CSP results with not only decision about the correctness of a composition with regard to the relation (5), but it also provides values for which the relation (5) is satisfied. In this way, we can obtain the concrete intervals (instead of just values) for all variables in all contracts (for correct compositions). This information can be useful for example when the composition M_{sys} has many alternative contracts, to observe which of them are identified as correct.

Relations or operations between variables in expressions are represented as constraints. Since both M_{sys} and CSP_{sys} use the SMT syntax for expressions, every operation is represented as a single constraint.

3.4.3 Components

From the perspective of structural organization, every component is represented in a CSP as a set of variables (inputs, outputs, parameters) from the integer domain, and a set of constraints, which correspond to the contracts implemented by that component (see Figure 4). Note that we distinguish here between variables used in

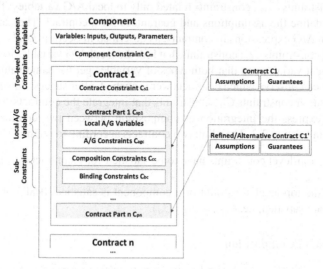

Fig. 4 Representation of a component in CSP (left) and an excerpt of the mapping the contracts to constraints and variables (right)

components, i.e. Σ in relation (4), and variables used in contracts, i.e. Σ in relation (2). Although they are identical, we define separated variables in the CSP for each of them. This means, when a component has two contracts, we have CSP variables for (a) component variables (inputs, outputs and parameters) and (b) CSP variables (inputs, outputs and parameters) for each contract. With this separation of contracts and components, we can identify which contracts are satisfied if the verification succeeds. As mentioned, the constraint solver not only responds with a decision, but it also finds all values of X_{CSP} for which the verification succeeds. Similarly, if the verification fails, the conflicting contracts can be easily identified.

Now we describe how the contracts are defined in a CSP, how they are linked with components, and how the criteria for correctness from relation (5) is represented in a CSP.

3.4.4 Contracts

As shown in Figure 4, each contract is represented as a single top-level constraint C_s. This constraint is further related to a set of local A/G variables (inputs, outputs, parameters) and a set of sub-constraints. The sub-constraints represent the constraints of the refined/abstracted or alternative contracts (contract parts Cs_p in figure). Because refined/abstracted and alternative contracts do not depend on each other, we define the top-level constraint C_s as follows: $C_s := (\vee_{i \in \mathbb{N}} Cs_{pi})$. In this relation, any contract which can satisfy the relation (5) implies that the top-level contract constraint C_s is satisfied.

As illustrated in Figure 4, every contract consists of the following sub-constraints:

- A/G constraints C_{agc}: constraints related only to local A/G variables. These constraints define the assumptions and guarantees for a contract. They are defined based on A/G expressions in contracts, as described in Section 3.4.2.
- Binding constraints C_{bc}: constraints that link the local A/G variables to the global component variables so that both types of variables get the same values. In this way, we can observe which contracts were satisfied, after successful verification.
- Composition constraints C_{cc}: constraints that integrate the contracts. These constraints express the integration or composition between two contracts, as described in Section 3.3. They link two contracts according to relation (5).

All three top-level constraints have to be satisfied for a contract C_{sp}, i.e. $C_{sp} := (C_{agc} \wedge C_{bc} \wedge C_{cc})$.

Finally, the top-level constraint of a component is satisfied, if all contract constraints C_s are satisfied, i.e. $C_m := (\wedge_{i \in \mathbb{N}} C_{si})$.

3.4.5 System/Composition

The compositions have very similar structure to basic or atomic components. Because they abstract some contracts of the contained components, additional constraints are defined to link these variables. An example of such a composition is given in Figure 3, where assumptions and guarantees of the contract C_{II} are an abstraction of assumptions and guarantees of the contained contracts.

Like atomic components, the complete component-based system M_{sys} is represented in a CSP as a set of variables and constraints. Within this set of constraints, there is a single top-level constraint of the composition C_m which links the complete hierarchy of the sub-constraints and variables discussed previously. The CSP has a solution only if this top-level constraint is satisfied. Finally, the C_m corresponds to the top-level constraint in the constraint set C_{CSP} from the relation (6).

4 Experimental Results

In the following, we describe the results of the preliminary evaluation and we discuss the performance of our approach.

To conduct the experiment, we used Java-based Choco constraint solver (choco Team (2010)). In our experiment, we defined the composition M_{sys} as a XML description, which is then used to generate the CSP in memory.

The main goal of this experiment is to show whether the proposed CSP is applicable to solve the composition problems defined with data properties, and for which system configurations. We conduct the experiment by showing how the verification responds with regard to attributes that might have an effect on runtime. These attributes include:

- Components and properties: how the verification scales with regard to number of components and properties, including also the presence of the alternative properties.
- Nature of properties: different properties may require different expressions in the CSP, including operations on fixed values, intervals, or more advanced operations such as ones used to define resource constraints (e.g. sum, min, etc.).

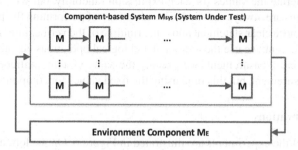

Fig. 5 System configuration used to conduct the experiments (M - component, M_E - environment component)

Figure 5 shows the system configuration used to conduct the experiments. The inputs for the verification are provided by the Environment component, which encloses the component-based system under test. All experiments were executed on Intel i7-3630QM, 4 cores, 2.40GHz.

4.1 Quantitative Results

For this experiment, we performed two measurements. In the first measurement, we show the response time with regard to the number of components, properties and alternative properties, having specified assumptions and guarantees as intervals. Then, in the second measurement, we use the same configurations but with fixed values for expressions. With these two measurements, we are able to observe the limits on modeling the component-based system with regard to number components, properties, and expressions used to describe the properties.

4.1.1 Measurements

In the first measurement, we execute several thousands of system configurations with the varying number of components and properties. The measurement has two parts. In the first part, we verify the system configurations with the varying number of components, each having varying number of properties but with constant number of assumptions or guarantees (i.e., each component variable is therefore related to only one expression). In the second part, each of the components has varying number of alternative and refined properties, so that many solutions are possible. In this case, each component variable is related to many expressions.

The expressions in the first measurement are defined in a way that always the intervals of the component variables have to be satisfied, and not the fixed values. An example for such expression is given in Section 3.2 for the contract C_{II}, which is satisfied only if the variables s_{en} and a_{tr} are in ranges $[0, 6400]$ and $[0, 100]$ respectively.

For the input test data, i.e. the operands of the assumption and guarantee expressions, we generate the values for each expression randomly, but with the rule that the assumptions are always satisfied. The advantage of performing the positive tests here is to get more clear statement about the runtime of the verification. In both parts of the measurement, we use the relational and logical operations on values.

In the second measurement, we execute the same system configurations as in previous measurement, but this time using the fixed values for component variables.

4.1.2 Observations

First results of the experiments are illustrated in Figure 6. On the left, an excerpt of the results for the first measurement is shown, where the properties have a constant number of assumptions and guarantees. The reason why the verification responds in

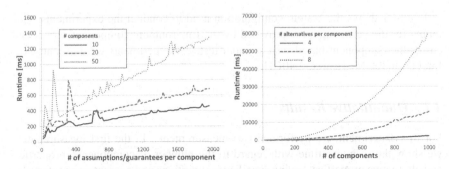

Fig. 6 Experimental results: runtime for system configuration with varying number of assumption/guarantee expressions and components (left) and varying number of components and alternative properties (right)

short time is that each component variable has only one expression (assumption or guarantee constraint, C_{agc}), and it is then immediately instantiated to a value indicated by that expression. The runtime depends in this case therefore on the number of components and properties.

On the right in Figure 6, a scenario that is more likely to occur in practice is shown. Here, each component variable has an increasing number of expressions, and these expressions are alternatives (as mentioned in the description of the measurement). The response time of the verification strongly depends on the number of alternatives, because each of the expressions represents different interval. The solver has to adjust the component variables to adequate intervals, in order to find a solution. Furthermore, since the choice of the particular alternative may influence the choice of the intervals in other connected components, often the backtracks have to be done to the state where the constraints were satisfied, which is time consuming.

In the second measurement, we observed very similar results as illustrated in Figure 6 on the left. Having fixed values on component variables, no search has to be performed, but just the constraint verification. For the case where the alternatives are used, more time is required to find a solution, but this time is negligible in contrast to situation when using intervals (i.e. Figure 6, right).

In the end, we summarize our observations with Figure 7. This figure shows the region for which the verification can complete in a "reasonable time". We set the limit for this time on 2 minutes, just to get a first feedback about possible configurations for the system under test. To establish this region, we used the system configuration with the worst case in response time, i.e. the one having the alternative properties from the first measurement.

4.2 Qualitative Results: Discussion

Figure 7 shows the worst-case scenario, in which a component-based system is modelled having varying number of assume guarantee expressions. The verification

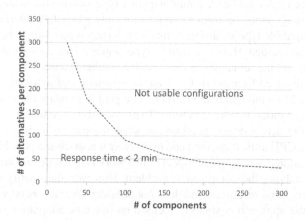

Fig. 7 Region of possible system configurations for which the verification completes within a given time

scales well but for configurations with only few instances of either components or properties. In nowadays automotive systems for example, there are more than 800 software components, that control various technical sub-processes in automobiles

(Kindel and Friedrich (2009)). However, it is still possible to support these configurations, since each such sub-system can be provided to verification independently,
and also, not all components are massively interconnected as in Figure 5. For example, the simplified system from Figure 1 is modelled using 13 software components
(is just one option to realize that system).

5   Related Work

Now we turn to a brief overview of related studies. We summarize here some relevant articles that handle compositional verification based on data semantics.

Similar problems to those described in our problem statement were identified by
Sun et. al (Sun et al (2009)) in their work on verifying the composition of analogue
circuits for analogue system design. In their approach, each analogue element (re-
sistor, capacitor, etc.) is characterized by its performance profile and this profile is
used to build the contract; that is, for certain values of the inputs the element responds with certain output values. Using contracts made from performance profiles,
it was possible to eliminate many integration failures early in the system design
phase. These structural compositions of analogue elements are very similar to the
compositions in CBSE. However, the model of Sun et al. only considers connections
between elements (horizontal relations).

Another article describes a runtime framework for dynamic adaptation of safety-
critical systems in the automotive domain (Adler et al (2011)). In the event of failures or degradation of quality, the intent is to reconfigure the automotive system
while it is operating. In contrast to the previous approach, the compositional verification in this case is based on a common quality type system shared among components. Two components can form a composition only when their interfaces or ports
have the compatible type qualities. In this way, wrong type castings between components can be avoided. However, using a type system in our case would just verify
the syntax but not the semantics of data (i.e. the concrete values).

A more advanced framework for dynamic adaptation of avionics systems was
developed by Montano (Montano (2011)). The goal is to adapt the system to new,
correct configurations, in case of failures. To perform this, a common quality system
defines the contracts between functions and available static resources (e.g. memory
consumption, CPU utilization, etc.) and in this way it restricts the possible set of correct configurations. An important aspect of this work is that it demonstrates the CSP
approach to solving the composition problem. However, the quality type system
only considers static resources, and does not consider contracts between functions.
Ultimately, the approach is strongly focused on dynamic adaptation with human-
assisted decision making.

In the field of industrial automation, the authors in (de Sousa (2012)) propose
the static verification of compositions based on data types of the IEC 61131-3
component model (or standard). This model defines the standard data types but it
also allows definition of customized data types (derived from existing ones) and
combination of existing data types into complex structures. The authors identified

ambiguities in the standard for user-defined data types and defined a proper compatibility criteria. Like the adaptation approach in the automotive domain (Adler et al (2011)), this work considers only a type system. However, the approach verifies not only compositions, but also the use of variables in IEC 61131-related languages.

In the last few years, several research projects have begun to handle the topics of compositional verification (SPEEDS (2006-2012)), (COMPASS (2011-2014)), (SAFECER (2011-2015)) by formalizing system models (component models) and languages for specification of contracts. These approaches share many concepts, especially contract-based design and formal behavioural verification of compositions. Although our model is conceptually very similar, it differs in that it considers the data semantics of property values, and it addresses a specific type of component-based systems in which data semantics can be used to express the validity criteria for compositions.

6 Conclusion

In this paper, we presented a method for the verification of compositions in component-based systems. The components modelled here are enriched with properties, which describe the data semantics of components. The novelty of our verification lies in representing the composition along with modelled properties as a Constraint Satisfaction Problem (CSP), which allows us to achieve two important objectives. First, using relational, logical and more advanced operators on data, many types of properties can be supported. Second, for properties that use basic logical and arithmetic operators, the verification can scale up to several hundreds of components, each of them consisting of few tens of properties, which makes the approach promising for the use in practice.

As part of our ongoing work, we want to characterize the runtime performance based on different types of properties, since they impact the scalability at most. In addition, we also want to investigate other parameters such as solver search policy, solver engine, etc., in order to find best configuration for the verification method.

References

Adler, R., Schaefer, I., Trapp, M., Poetzsch-Heffter, A.: Component-based modeling and verification of dynamic adaptation in safety-critical embedded systems. ACM Trans. Embed. Comput. Syst. 10(2), 20:1–20:39 (2011),
http://doi.acm.org/10.1145/1880050.1880056,
doi:10.1145/1880050.1880056

de Alfaro, L., Henzinger, T.A.: Interface automata. SIGSOFT Softw. Eng. Notes 26(5), 109–120 (2001), http://doi.acm.org/10.1145/503271.503226,
doi:10.1145/503271.503226

Apt, K.: Principles of Constraint Programming. Cambridge University Press, New York (2003)

Basu, A., Bensalem, S., Bozga, M., Combaz, J., Jaber, M., Nguyen, T.H., Sifakis, J.: Rigorous component-based system design using the bip framework. IEEE Software 28(3), 41–48 (2011), doi:10.1109/MS.2011.27

Benveniste, A., Caillaud, B., Nickovic, D., Passerone, R., Raclet, J.B., Reinkemeier, P., Sangiovanni-Vincentelli, A., Damm, W., Henzinger, T., Larsen, K. (2012) Contracts for Systems Design. Tech. rep., Research Report, Nr. 8147, Inria (November 2012)

Butz, H.: (-) Open integrated modular avionic (ima): State of the art and future development road map at airbus deutschland. Department of Avionic Systems at Airbus Deutschland GmbH Kreetslag 10, D-21129 Hamburg, Germany

choco Team, choco: an Open Source Java Constraint Programming Library. Research report 10-02-INFO, École des Mines de Nantes (2010)

Earle, C.B., Gómez-Martínez, E., Tonetta, S., Puri, S., Mazzini, S., Gilbert, J.L., Hachet, O., Oliver, R.S., Ekelin, C., Zedda, K.: Languages for Safety-Certification Related Properties. In: Proc. Work in Progress Session at 39th Euromicro Conf. on Software Engineering and Advanced Applications (SEAA 2013) (2013)

COMPASS (2011-2014) Compass - comprehensive modelling for advanced systems of systems, http://www.compass-research.eu

Crnkovic, I.: Building Reliable Component-Based Software Systems. Artech House, Inc., Norwood (2002)

Frey, P.: Case Study: Engine Control Application. Tech. rep., Ulmer Informatik-Berichte, Nr. 2010-03 (2010)

Gössler, G., Sifakis, J.: Composition for component-based modeling. Sci. Comput. Program 55(1-3), 161–183 (2005),
http://dx.doi.org/10.1016/j.scico.2004.05.014,
doi:10.1016/j.scico.2004.05.014

Kindel, O., Friedrich, M.: Softwareentwicklung mit AUTOSAR: Grundlagen, Engineering, Management in der Praxis. dpunkt Verlag; Auflage: 1 (Juni 8, 2009)

Montano, G.: Dynamic reconfiguration of safety-critical systems: Automation and human involvement. PhD Thesis (2011)

SAFECER (2011-2015) Safecer - safety certification of software-intensive systems with reusable components, http://safecer.eu

Sentilles, S., Štěpán, P., Carlson, J., Crnković, I.: Integration of extra-functional properties in component models. In: Lewis, G.A., Poernomo, I., Hofmeister, C. (eds.) CBSE 2009. LNCS, vol. 5582, pp. 173–190. Springer, Heidelberg (2009),
http://dx.doi.org/10.1007/978-3-642-02414-6_11

de Sousa, M.: Data-type checking of iec61131-3 st and il applications. In: 2012 IEEE 17th Conference on Emerging Technologies Factory Automation (ETFA), pp. 1–8 (2012), doi:10.1109/ETFA.2012.6489534

SPEEDS (2006-2012) Speculative and exploratory design in systems engineering - speeds, http://www.speeds.eu.com

Sun, X., Nuzzo, P., Wu, C.C., Sangiovanni-Vincentelli, A.: Contract-based system-level composition of analog circuits. In: 46th ACM/IEEE Design Automation Conference, DAC 2009, pp. 605–610. Los Alamitos (2009)

Tran, E.: Verification/validation/certification. Carnegie Mellon University, 18-849b Dependable Embedded Systems (1999)

A Polynomial-Time Algorithm for Checking the Equivalence of Deterministic Restricted One-Counter Transducers Which Accept by Final State

Mitsuo Wakatsuki, Etsuji Tomita, and Tetsuro Nishino

Abstract. This paper is concerned with a subclass of deterministic pushdown transducers, called deterministic restricted one-counter transducers (droct's), and studies the equivalence problem for droct's which accept by final state. In the previous study, we presented a polynomial-time algorithm for checking the equivalence of real-time droct's. By extending the technique, we present a polynomial-time algorithm for checking the equivalence of non-real-time droct's.

Keywords: formal language theory, equivalence problem, deterministic pushdown transducer, deterministic restricted one-counter transducer, polynomial-time algorithm.

1 Introduction

One of the most interesting questions in formal language theory is the equivalence problem for deterministic pushdown automata (dpda's) and the corresponding deterministic context-free grammars. Although Sénizergues [3] has proved that the equivalence problem for any pair of dpda's is decidable, his algorithm is very complicated and is hard to be implemented. A checking the equivalence for some dpda's can play an important role in the learning process for these dpda's [4]. From a

Mitsuo Wakatsuki · Tetsuro Nishino
Graduate School of Informatics and Engineering, The University of
Electro-Communications, 1-5-1 Chofugaoka, Chofu, Tokyo 182-8585, Japan
e-mail: {wakatsuki.mitsuo,nishino}@uec.ac.jp

Etsuji Tomita
The Advanced Algorithms Research Laboratory, The University of
Electro-Communications, 1-5-1 Chofugaoka, Chofu, Tokyo 182-8585, Japan
e-mail: tomita@ice.uec.ac.jp

© Springer International Publishing Switzerland 2015   131
R. Lee (ed.), *SNPD*,
Studies in Computational Intelligence 569, DOI: 10.1007/978-3-319-10389-1_10

practical point of view, it is desirable to have a polynomial-time algorithm for checking the equivalence.

A *deterministic one-counter automaton* (*doca*) is a pdda having only one stack symbol, with the exception of a bottom-of-stack marker. A *deterministic restricted one-counter automaton* (*droca*) is a dpda which has only one stack symbol. The class of languages accepted by droca's is a proper subclass of that of languages accepted by doca's. Moreover, the class of languages accepted by droca's which accept by final state properly contains the class of regular languages. Valiant has shown that the equivalence problem for doca's is decidable in single exponential time [8] [9] and the inclusion problem for doca's is undecidable [8]. On the other hand, Higuchi et al. have presented polynomial-time algorithms for checking the inclusion (also the equivalence) of droca's which accept by empty stack [1] and of real-time droca's which accept by final state [2].

A deterministic pushdown transducer (dpdt) is a dpda provided with outputs. The equivalence problem for dpdt's is essentially more difficult than that for dpda's. In this paper, we are concerned with a subclass of dpdt's called *deterministic restricted one-counter transducers* (*droct's*), which are droca's provided with outputs, and study the equivalence problem for non-real-time droct's which accept by final state. Since these droct's may have infinite sequences of ε-moves, it is possible that their stack heights increase infinitely without reading inputs. In the previous study, we presented a polynomial-time algorithm for checking the equivalence of a pair of real-time droct's (i.e. droct's without ε-moves) [10]. By extending the technique in Ref. [10], we present a new direct branching algorithm for checking the equivalence of non-real-time droct's (i.e. droct's with possible ε-moves). The worst-case time complexity of our algorithm is polynomial in the description length of these droct's.

2 Definitions and Notation

We assume that the reader is familiar with the basics of automata and formal language theory. Our definitions and notation are almost as in Ref. [10].

Definition 1. A *deterministic pushdown transducer* (*dpdt* for short) which accepts by final state is denoted by $T = (Q, \Gamma, \Sigma, \Delta, \mu, q_0, Z_0, F)$, where $Q, \Gamma, \Sigma, \Delta, \mu$ are the finite sets of states, stack symbols, input symbols, output symbols, and transition-output rules respectively, $q_0 \ (\in Q)$ is the initial state, $Z_0 \ (\in \Gamma)$ is the initial stack symbol, and $F \ (\subseteq Q)$ is the set of final states. We denote an empty string in Γ^*, Σ^* or Δ^* by ε.

The set μ of *transition-output rules* is a set of rules of the form $(p, A) \xrightarrow{a/z} (q, \theta)$ with $p, q \in Q$, $A \in \Gamma$, $a \in \Sigma \cup \{\varepsilon\}$, $z \in \Delta^*$, $\theta \in \Gamma^*$, that satisfies the following conditions:

(i) If $(p, A) \xrightarrow{a/z} (q, \theta)$ with $a \in \Sigma \cup \{\varepsilon\}$ is in μ, then μ contains no rule of the form $(p, A) \xrightarrow{a/z'} (r, \gamma)$ for any $(r, \gamma) \neq (q, \theta)$, $z' \neq z$.

(ii) If $(p, A) \xrightarrow{\varepsilon/z} (q, \theta)$ is in μ, then μ contains no rule of the form $(p, A) \xrightarrow{a/z'} (r, \gamma)$

for any $a \in \Sigma$, $z' \in \Delta^*$, $r \in Q$, $\gamma \in \Gamma^*$. Such a rule as $(p,A) \xrightarrow{\varepsilon/z} (q,\theta)$ is called an ε-rule.

The *deterministic pushdown automaton* (*dpda* for short) $M = (Q, \Gamma, \Sigma, \delta, q_0, Z_0, F)$ with

$$\delta = \{(p,A) \xrightarrow{a} (q,\theta) \mid (p,A) \xrightarrow{a/z} (q,\theta) \in \mu, z \in \Delta^*\}$$

is called the *associated* dpda for the above dpdt T, and is just as in Ref. [5], Definition 2.1, pp.190–191. A dpdt or a dpda is said to be *real-time* if it has no ε-rules.

Definition 2. A dpda M (respectively, a dpdt T) is said to be a *deterministic restricted one-counter automaton* (*transducer*, resp.) (*droca* (*droct*, resp.) for short) if $\Gamma = \{Z_0\}$. When the droca M is the associated dpda for the droct T, M is called the *associated droca* for T.

Definition 3. A *configuration* (p, α) of the dpdt T, with its associated dpda M, is an element of $Q \times \Gamma^*$, where the *leftmost* symbol of α is the *top* symbol on the stack. In particular, (q_0, Z_0) is called the *initial configuration*.

A configuration (p, α) is said to be in *reading mode* if $\alpha = A\alpha'' \in \Gamma^+$ and $(p,A) \xrightarrow{a/z} (q,\theta) \in \mu$ for some $a \in \Sigma$, $z \in \Delta^*$ and $(q,\theta) \in Q \times \Gamma^*$, while it is said to be in ε-*mode* if $\alpha = A\alpha'' \in \Gamma^+$ and $(p,A) \xrightarrow{\varepsilon/z} (q,\theta) \in \mu$ for some $z \in \Delta^*$ and $(q,\theta) \in Q \times \Gamma^*$.

The *height* of a configuration (p, α) is $|\alpha|$. Here, for a string α, $|\alpha|$ denotes the length of α. Moreover, for a set S, $|S|$ denotes the cardinality of S.

Definition 4. The dpdt T, with its associated dpda M, makes a *move*

$$(p, A\omega) \xrightarrow[T]{a/z} (q, \theta\omega)$$

for any $\omega \in \Gamma^*$ iff μ contains a rule $(p,A) \xrightarrow{a/z} (q,\theta)$ with $a \in \Sigma \cup \{\varepsilon\}$. A sequence of such moves through successive configurations as

$$(p_i, \alpha'_i \alpha'') \xrightarrow[T]{a_i/z_i} (p_{i+1}, \alpha'_{i+1}\alpha''), \ 1 \le i \le m,$$

is called a *derivation*, and is written as

$$(p_1, \alpha'_1 \mid \alpha'') \xrightarrow[T]{x/y} {}^{(m)} (p_{m+1}, \alpha'_{m+1} \mid \alpha''),$$

where $x = a_1 a_2 \cdots a_m$ and $y = z_1 z_2 \cdots z_m$, or simply

$$(p_1, \alpha'_1 \alpha'') \xrightarrow[T]{x/y} (p_{m+1}, \alpha'_{m+1}\alpha''),$$

if $(p_1, \alpha'_1 \mid \alpha'') \xrightarrow[M]{x} {}^{(m)} (p_{m+1}, \alpha'_{m+1} \mid \alpha'')$ as in Ref. [5], Definition 2.3, pp.191–192. By convention, we let $(p, \alpha) \xrightarrow[T]{\varepsilon/\varepsilon} (p, \alpha)$ for any $(p, \alpha) \in Q \times \Gamma^*$.

A derivation $(p, \alpha) \xoverset{x/y}{\underset{T}{\Longrightarrow}} (q, \beta)$ is also written as $(p, \alpha) \xoverset{x/y}{\underset{T}{\Longrightarrow}\!\!\!\gg} (q, \beta)$ if no such derivation as $(q, \beta) \xoverset{\varepsilon/z}{\underset{T}{\Longrightarrow}} (r, \gamma)$ with $z \in \Delta^*$ is possible for any $(r, \gamma) \neq (q, \beta)$. Moreover, a derivation $(p, \alpha) \xoverset{x/y}{\underset{T}{\Longrightarrow}} (q, \beta)$ is also written as $(p, \alpha) \xoverset{x/y}{\underset{T}{\Longrightarrow}}\!|_F (q, \beta)$ if no such derivation as $(p, \alpha) \xoverset{x/y'}{\underset{T}{\Longrightarrow}} (r, \gamma) \xoverset{\varepsilon/y''}{\underset{T}{\Longrightarrow}} (q, \beta)$ with $y'y'' = y$ is possible for any $(r, \gamma) \in F \times \Gamma^*$ such that $(r, \gamma) \neq (q, \beta)$ and no such derivation as $(q, \beta) \xoverset{\varepsilon/z}{\underset{T}{\Longrightarrow}} (s, \eta)$ with $q \notin F, z \in \Delta^*$ is possible for any $(s, \eta) \in Q \times \Gamma^*$ such that $(s, \eta) \neq (q, \beta)$.

Definition 5. For a configuration $(p, \alpha) \in Q \times \Gamma^*$ of the dpdt T which accepts by final state, define

$$N(p, \alpha) = \{x \in \Sigma^* \mid (p, \alpha) \xoverset{x/y}{\underset{T}{\Longrightarrow}} (q, \varepsilon) \text{ for some } q \in Q, y \in \Delta^*\},$$

$$L(p, \alpha) = \{x \in \Sigma^* \mid (p, \alpha) \xoverset{x/y}{\underset{T}{\Longrightarrow}} (q, \beta) \text{ for some } q \in F, \beta \in \Gamma^*, y \in \Delta^*\}$$

and

$$\text{TRANS}(p, \alpha) = \{x/y \in \Sigma^* \times \Delta^* \mid (p, \alpha) \xoverset{x/y}{\underset{T}{\Longrightarrow}} (q, \beta) \text{ for some } q \in F, \beta \in \Gamma^*\}.$$

A configuration (p, α) is said to be *live* if $L(p, \alpha) \neq \emptyset$. Moreover, define $N(T) = N(q_0, Z_0)$. The *language* accepted by T is defined to be $L(T) = L(q_0, Z_0)$, and the *translation* defined by T is $\text{TRANS}(T) = \text{TRANS}(q_0, Z_0)$.

Definition 6. For a configuration (p, α) of the dpdt T which accepts by final state, let $(p, \alpha) \xoverset{\varepsilon/z}{\underset{T}{\Longrightarrow}\!\!\!\gg} (p', \alpha')$ for some $z \in \Delta^*$. Then, if $\alpha' = A'\alpha'' \neq \varepsilon$ with $A' \in \Gamma$, define

$$\text{FIRST}(p, \alpha) = \text{FIRST}(p', A')$$
$$= \{a \in \Sigma \mid (p', A') \xoverset{a/z'}{\underset{}{\longrightarrow}} (q, \theta) \in \mu \text{ for some } (q, \theta) \in Q \times \Gamma^*, z' \in \Delta^*\}.$$

Otherwise, define $\text{FIRST}(p, \alpha) = \emptyset$.

Moreover, let $(p, \alpha) \xoverset{\varepsilon/z}{\underset{T}{\Longrightarrow}}\!|_F (p', \alpha')$ for some $z \in \Delta^*$. Then, if $p' \notin F$, define

$$\text{FIRST}_{live}(p, \alpha) = \{a \in \text{FIRST}(p', \alpha') \mid$$
$$(p', \alpha') \xoverset{a/z'}{\underset{T}{\Longrightarrow}} (q, \beta), L(q, \beta) \neq \emptyset \text{ for some } (q, \beta) \in Q \times \Gamma^*, z' \in \Delta^*\}.$$

Otherwise (i.e. $p' \in F$), define

$$\text{FIRST}_{live}(p, \alpha) = \{\varepsilon\} \cup \{a \in \text{FIRST}(p', \alpha') \mid$$
$$(p', \alpha') \xoverset{a/z'}{\underset{T}{\Longrightarrow}} (q, \beta), L(q, \beta) \neq \emptyset \text{ for some } (q, \beta) \in Q \times \Gamma^*, z' \in \Delta^*\}.$$

Definition 7. For $w, h, t \in \Delta^*$, let $h^{-1}w = t$ if $w = ht$, and $wt^{-1} = h$ if $w = ht$. Then for Δ, let $\Delta^{-*} = \{h^{-1} \mid h \in \Delta^*\}$, $\Delta^{\pm*} = \Delta^* \cup \Delta^{-*}$, where $\Delta^* \cap \Delta^{-*} = \{\varepsilon\}$, and for $h \in \Delta^{\pm*}$, let $(h^{-1})^{-1} = h$. Moreover, let $\Delta^{-+} = \Delta^{-*} - \{\varepsilon\}$.

For $k^{-1} \in \Delta^{-*}$ with $k \in \Delta^*$, define $|k^{-1}| = -|k|$. Then for $h \in \Delta^{\pm*}$, define

$$||h|| = \begin{cases} |h| & \text{if } h \in \Delta^*, \\ |k| & \text{if } h = k^{-1} \in \Delta^{-*}. \end{cases}$$

Definition 8. Let (p, α) be a configuration of a dpdt T_1 which accepts by final state, $(\bar{p}, \beta)$ be that of a dpdt T_2 which accepts by final state, and $h \in \Delta^{\pm*}$. If $\text{TRANS}(p, \alpha) = h\,\text{TRANS}(\bar{p}, \beta) = \{x/hv \mid x/v \in \text{TRANS}(\bar{p}, \beta)\}$, then it is written as $(p, \alpha) \equiv h(\bar{p}, \beta)$. Here, if $h \in \Delta^{-*}$, then $(p, \alpha) \equiv h(\bar{p}, \beta)$ is the same as $k(p, \alpha) \equiv (\bar{p}, \beta)$ with $k = h^{-1} \in \Delta^*$ which stands for

$$\{x/ku \mid x/u \in \text{TRANS}(p, \alpha)\} = \text{TRANS}(\bar{p}, \beta).$$

Such a formula as $(p, \alpha) \equiv h(\bar{p}, \beta)$, $h \in \Delta^{\pm*}$ or $k(p, \alpha) \equiv (\bar{p}, \beta)$, $k \in \Delta^*$, is called an *equivalence equation*.

If $\text{TRANS}(T_1) = \text{TRANS}(T_2)$, then the two dpdt's are *equivalent*, and it is written as $T_1 \equiv T_2$. Otherwise, $T_1 \not\equiv T_2$.

3 Basic Properties and Propositions

For a droct $T = (Q, \Gamma, \Sigma, \Delta, \mu, q_0, Z_0, F)$, define $\tau = \text{Max}\{|z| \mid (p, A) \xrightarrow{a/z} (q, \theta) \in \mu\}$ and $\rho = \text{Max}\{|\theta| \mid (p, A) \xrightarrow{a/z} (q, \theta) \in \mu\}$. Without loss of generality, we may assume that $\rho \leq 2$, i.e. stack height increases by at most one per one move.

3.1 Basic Properties of DROCT's Which Accept by Final State

Lemma 1. *Let* $T = (Q, \Gamma, \Sigma, \Delta, \mu, q_0, Z_0, F)$ *be a droct which accepts by final state, and* $(p, \alpha) \in Q \times \Gamma^*$ *be a configuration of* T. *Then, the followings (i) – (iii) hold.*
(i) If $N(p, \alpha) \neq \emptyset$ *with* $|\alpha| \geq |Q|$, *then* $N(p, \beta) \neq \emptyset$ *for any* $\beta \in \Gamma^*$.
(ii) For any $\beta \in \Gamma^*$, *it holds that*
(a) $L(p, \alpha) \subseteq L(p, \alpha\beta)$ *and*
(b) $\text{TRANS}(p, \alpha) \subseteq \text{TRANS}(p, \alpha\beta)$.
(iii) If $N(p, \alpha) = \emptyset$, *then for any* $\beta \in \Gamma^*$, *it holds that*
(a) $N(p, \alpha\beta) = \emptyset$,
(b) $L(p, \alpha) = L(p, \alpha\beta)$ *and*
(c) $\text{TRANS}(p, \alpha) = \text{TRANS}(p, \alpha\beta)$.

Proof. (i): It is clear because $\Gamma = \{Z_0\}$ (See Ref. [1], Lemma 3.1, p.306 for the details). (ii), (iii): It follows from Definition 5 since T is the dpdt which accepts by final state. $\qquad\square$

From Lemma 1 (i) and (iii)(a), we can construct a checking procedure whether any configuration $(p, \alpha) \in Q \times \Gamma^*$ of a given droct T is $N(p, \alpha) \neq \emptyset$ or not since it suffices to examine all configurations whose height is at most $|Q| + 1$. Then, the worst-case time complexity of this procedure is polynomial with respect to the description length of T.

Lemma 2. *Let* $T = (Q, \Gamma, \Sigma, \Delta, \mu, q_0, Z_0, F)$ *be a droct which accepts by final state, and* $(p, \alpha) \in Q \times \Gamma^*$ *be a configuration of* T. *If there exists a derivation such that* $(p, \alpha) \xrightarrow[T]{x/y} (q, \beta)$ *for some* $x \in \Sigma^*, y \in \Delta^*, (q, \beta) \in Q \times \Gamma^*$, *then there exists a derivation such that* $(p, \alpha) \xrightarrow[T]{x'/y'}{}^{(n)} (q, \beta')$ *with* $n \leq |Q|(|Q| - 1)$ *for some* $x' \in \Sigma^*, y' \in \Delta^*, \beta' \in \Gamma^*$.

Proof. It is clear because $\Gamma = \{Z_0\}$ (See Ref. [2], Lemma 3.1, p.941 for the details). □

From Lemma 2, we can construct a checking procedure whether any configuration $(p, \alpha) \in Q \times \Gamma^*$ of T is *live* or not since for all configurations $(r, \gamma) \in Q \times \Gamma^*$ such that $(p, \alpha) \xrightarrow[T]{x/y}{}^{(n)} (r, \gamma)$ with $n \leq |Q|(|Q| - 1)$ for some $x \in \Sigma^*$ and $y \in \Delta^*$, it suffices to check whether $r \in F$ or not. This procedure can be also performed in a polynomial-time with respect to the description length of T.

3.2 The Basic Proposition Concerning the Equivalence

We shall check the equivalence of two droct's $T_i = (Q_i, \Gamma_i, \Sigma, \Delta, \mu_i, q_{0i}, Z_{0i}, F_i)$ $(i = 1, 2)$ which accept by final state, whose associated droca's are M_i respectively. We are only concerned with the case where $L(T_i) \neq \emptyset$ $(i = 1, 2)$. Let $\tau = \text{Max}\{\tau_1, \tau_2\}$, where τ_i $(i = 1, 2)$ are constants depending on T_i defined as above.

To begin with, we give the most elementary proposition concerning outputs.

Proposition 1. *(i) Suppose* $T_1 \equiv T_2$ *holds and*

$$(q_{01}, Z_{01}) \xrightarrow[T_1]{w/w_1} (p, \alpha) \text{ with } L(p, \alpha) \neq \emptyset, \tag{1}$$

$$(q_{02}, Z_{02}) \xrightarrow[T_2]{w/w_2} (\bar{p}, \beta) \text{ with } L(\bar{p}, \beta) \neq \emptyset, \tag{2}$$

for some $w \in \Sigma^*, w_1, w_2 \in \Delta^*, p \in Q_1, \bar{p} \in Q_2, \alpha \in \Gamma_1^*$ *and* $\beta \in \Gamma_2^*$, *then it holds that*

$$w_1 h = w_2 \text{ for some } h \in \Delta^{\pm *} \tag{3}$$

and

$$(p, \alpha) \equiv h(\bar{p}, \beta). \tag{4}$$

(ii) If we have, in addition to (i), another pair of derivations $(q_{01}, Z_{01}) \xRightarrow[T_1]{w'/w'_1} (p, \alpha')$

with $L(p, \alpha') \neq \emptyset$ *and* $(q_{02}, Z_{02}) \xRightarrow[T_2]{w'/w'_2} (\bar{p}, \beta')$ *with* $L(\bar{p}, \beta') \neq \emptyset$ *for some* $w' \in \Sigma^*$, $w'_1, w'_2 \in \Delta^*$ *such that* $w'_1 h' = w'_2$ *for some* $h' \in \Delta^{\pm*}$, $\alpha' \in \Gamma_1^*$, $\beta' \in \Gamma_2^*$, *and hence* $(p, \alpha') \equiv h'(\bar{p}, \beta')$, *then* $h = h'$.

Proof. These properties can be easily derived from Definitions 5, 6, and 8, and Lemma 1 (ii). □

The above $h \in \Delta^{\pm*}$ in eq.(3) and eq.(4) is to compensate the difference between the two outputs, and is called an *output compensating part*. From Proposition 1 (ii), the output compensating part is unique for each equivalence equation as eq.(4) when $T_1 \equiv T_2$.

4 The Equivalence Checking Algorithm

The equivalence checking is carried out by developing step by step the so-called *comparison tree* as in Refs. [5]–[7].

At the initial stage, the comparison tree contains only the root labeled $(q_{01}, Z_{01}) \equiv (q_{02}, Z_{02})$ which is said to be in *unchecked* status. In each step, the algorithm considers a node labeled $(p, \alpha) \equiv h(\bar{p}, \beta)$ such that eq.(1) and eq.(2) with eq.(3) for some $w \in \Sigma^*$, $w_1, w_2 \in \Delta^*$, and tries to prove or disprove this equivalence. In the case where $\text{FIRST}_{live}(p, \alpha) = \text{FIRST}_{live}(\bar{p}, \beta) = \{\varepsilon\}$ and both (p, α) and $(\bar{p}, \beta)$ are not in ε-mode, if $h = \varepsilon$ then we turn the node to be in *checked* status. Otherwise, i.e. $h \neq \varepsilon$, conclude that "$T_1 \not\equiv T_2$". Also, if another internal node with the same label has appeared elsewhere in the tree, then we turn the above node to be *checked*. From Proposition 1 (ii), if another internal node labeled $(p, \alpha') \equiv h'(\bar{p}, \beta')$ such that $h' \neq h$ has appeared elsewhere in the tree, then conclude that "$T_1 \not\equiv T_2$". Otherwise, we expand it by *branching*, *skipping*, or *stack reduction*, which will be described below.

4.1 Branching

Branching is a basic step of developing the comparison tree. The following branching step is almost the same as in [10] but some extensions.

Lemma 3. *The equivalence equation eq.(4) holds iff the following conditions (i), (ii) and (iii) hold.*
(i) $\text{FIRST}_{live}(p, \alpha) = \text{FIRST}_{live}(\bar{p}, \beta)$.
(ii) (a) *In case both* (p, α) *and* $(\bar{p}, \beta)$ *are in reading mode, for each* $a_i \in \text{FIRST}_{live}(p, \alpha) - \{\varepsilon\} = \{a_1, a_2, \ldots, a_l\} \subseteq \Sigma$ $(i = 1, 2, \ldots, l)$, *let*

$$(p, \alpha) \xrightarrow[T_1]{a_i/z_i} (p_i, \alpha_i) \text{ and } (\bar{p}, \beta) \xrightarrow[T_2]{a_i/\bar{z}_i} (\bar{p}_i, \beta_i)$$

for some $(p_i, \alpha_i) \in Q_1 \times \Gamma_1^*$, $(\bar{p}_i, \beta_i) \in Q_2 \times \Gamma_2^*$, $z_i, \bar{z}_i \in \Delta^*$. *Then it holds that* $z_i h_i = h \bar{z}_i$ *for some* $h_i \in \Delta^{\pm *}$, $i = 1, 2, \ldots, l$.

(b) *In case* (p, α) *is in* ε-*mode, let* $a_1 = \varepsilon$, $l = 1$,

$$(p, \alpha) \xrightarrow[T_1]{\varepsilon / z_1} |_F (p_1, \alpha_1) \ \text{and} \ (\bar{p}, \beta) = (\bar{p}_1, \beta_1).$$

Then it holds that $z_1 h_1 = h$.

(c) *In case* (p, α) *is not in* ε-*mode and* $(\bar{p}, \beta)$ *is in* ε-*mode, let* $a_1 = \varepsilon$, $l = 1$,

$$(p, \alpha) = (p_1, \alpha_1) \ \text{and} \ (\bar{p}, \beta) \xrightarrow[T_2]{\varepsilon / \bar{z}_1} |_F (\bar{p}_1, \beta_1).$$

Then it holds that $h_1 = h \bar{z}_1$.

(iii) *Concerning the above condition* (ii), *it holds that* $(p_i, \alpha_i) \equiv h_i(\bar{p}_i, \beta_i)$, $i = 1, 2, \ldots, l$.

Proof. It follows from Proposition 1 (i). □

The checking whether conditions (i) and (ii) in Lemma 3 hold or not is named *output branch checking* to the node labeled eq.(4) in question. When it is verified to hold, the checking is said to be *successful*. Then we expand the above node to have l sons in *unchecked* status labeled by (iii), with edges labeled $z_i \backslash a_i / \bar{z}_i$, $i = 1, 2, \ldots, l$, and we turn the node in question to be *checked*. The step of developing the comparison tree in this way is named *branching* to the node in question. If condition (i) or (ii) does not hold, conclude that "$T_1 \not\equiv T_2$".

When the branching has been applied to the node in question, the number of sons of it is at most $|\Sigma|$.

4.2　Stack Reduction

From Lemma 1, for any node labeled $(p, \alpha) \equiv h(\bar{p}, \beta)$, if $N(p, \alpha) = \emptyset$, then there exists $\alpha' \in \Gamma_1^+$ such that $N(p, \alpha') = \emptyset$ and $\text{TRANS}(p, \alpha') = \text{TRANS}(p, \alpha)$ (i.e. $(p, \alpha') \equiv (p, \alpha)$) with $1 \le |\alpha'| \le |Q_1|$. Similarly, if $N(\bar{p}, \beta) = \emptyset$, there exists $\beta' \in \Gamma_2^+$ such that $N(\bar{p}, \beta') = \emptyset$ and $\text{TRANS}(\bar{p}, \beta') = \text{TRANS}(\bar{p}, \beta)$ (i.e. $(\bar{p}, \beta') \equiv (\bar{p}, \beta)$) with $1 \le |\beta'| \le |Q_2|$.

Definition 9. Let $(p, \alpha) \equiv h(\bar{p}, \beta)$ be a label of the node in question. If $N(p, \alpha) = \emptyset$, let α_0 be the *shortest* stack such that $N(p, \alpha_0) = \emptyset$. Otherwise, let $\alpha_0 = \alpha$. Moreover, if $N(\bar{p}, \beta) = \emptyset$, let β_0 be the *shortest* stack such that $N(\bar{p}, \beta_0) = \emptyset$. Otherwise, let $\beta_0 = \beta$. If it holds that $|\alpha_0| < |\alpha|$ or $|\beta_0| < |\beta|$, we say that the stack reduction is *applicable* to the node in question.

When stack reduction is applicable to the node in question, we expand it to have only one son labeled by $(p, \alpha_0) \equiv h(\bar{p}, \beta_0)$ in *unchecked* status, with edge labeled $\varepsilon \backslash \lambda / \varepsilon$, where λ is a new symbol not in Σ^*, and then we turn the node in question to be *checked*. The step of developing the comparison tree in this way is named *stack reduction* to the node in question.

When the stack reduction has applied to the node in question, the following (i) or (ii) holds:

(i) $1 \leq |\alpha_0| \leq |Q_1|,\ |\alpha_0| < |\alpha|$ and $|\beta_0| \leq |\beta|$,
(ii) $|\alpha_0| \leq |\alpha|,\ 1 \leq |\beta_0| \leq |Q_2|$ and $|\beta_0| < |\beta|$.

4.3 Skipping

In order to prevent the comparison tree from growing larger and larger infinitely by successive application of branching or stack reduction steps, certain nodes are expanded by other steps of skipping. Let the comparison tree which has just been constructed up to a certain stage be denoted by $T(T_1 : T_2)$.

Definition 10. If $(p_1, \alpha_1 \gamma_1) \equiv h_1(\bar{p}_1, \bar{\alpha}_1 \bar{\gamma}_1)$ and $(p_2, \alpha_2 \gamma_1) \equiv h_2(\bar{p}_2, \bar{\alpha}_2 \bar{\gamma}_1)$ are labels of two nodes in $T(T_1 : T_2)$ which are connected by an edge labeled $u_1 \backslash x_1 / v_1 \in \Delta^* \times (\Sigma \cup \{\lambda\})^* \times \Delta^*$ such that

$$(p_1, \alpha_1) \xrightarrow[T_1]{\sigma(x_1)/u_1} (p_2, \beta_2) \text{ and } (\bar{p}_1, \bar{\alpha}_1) \xrightarrow[T_2]{\sigma(x_1)/v_1} (\bar{p}_2, \bar{\beta}_2),$$

with $u_1 h_2 = h_1 v_1$, $|\beta_2| \geq |\alpha_2|$, and $|\bar{\beta}_2| \geq |\bar{\alpha}_2|$, where $\sigma(x_1) \in \Sigma^*$ is the string obtained by eliminating all λ's from x_1, then we write

$$\langle (p_1, \alpha_1 \mid \gamma_1) \equiv h_1(\bar{p}_1, \bar{\alpha}_1 \mid \bar{\gamma}_1) \rangle \xrightarrow[T(T_1:T_2)]{u_1 \backslash x_1 / v_1}$$
$$\langle (p_2, \alpha_2 \mid \gamma_1) \equiv h_2(\bar{p}_2, \bar{\alpha}_2 \mid \bar{\gamma}_1) \rangle.$$

A sequence of such father-son relations as

$$\langle (p_i, \alpha_i \mid \gamma_1) \equiv h_i(\bar{p}_i, \bar{\alpha}_i \mid \bar{\gamma}_1) \rangle \xrightarrow[T(T_1:T_2)]{u_i \backslash x_i / v_i}$$
$$\langle (p_{i+1}, \alpha_{i+1} \mid \gamma_1) \equiv h_{i+1}(\bar{p}_{i+1}, \bar{\alpha}_{i+1} \mid \bar{\gamma}_1) \rangle$$

with $u_i h_{i+1} = h_i v_i$, for $i = 1, 2, \ldots, m$, is named a *derivation path*, and is written as

$$\langle (p_1, \alpha_1 \mid \gamma_1) \equiv h_1(\bar{p}_1, \bar{\alpha}_1 \mid \bar{\gamma}_1) \rangle \xrightarrow[T(T_1:T_2)]{u \backslash x / v}$$
$$\langle (p_{m+1}, \alpha_{m+1} \mid \gamma_1) \equiv h_{m+1}(\bar{p}_{m+1}, \bar{\alpha}_{m+1} \mid \bar{\gamma}_1) \rangle,$$

where $x = x_1 x_2 \cdots x_m$, $u = u_1 u_2 \cdots u_m$, and $v = v_1 v_2 \cdots v_m$. Here, "$\mid$" may be omitted.

Definition 11. Suppose that a node in question is labeled

$$(p, \omega_1 \alpha'') \equiv h(\bar{p}, \omega_2 \beta''), \tag{5}$$

where $\alpha = \omega_1 \alpha''$ and $\beta = \omega_2 \beta''$ with $\alpha'' \neq \varepsilon$ or $\beta'' \neq \varepsilon$. We say that *the prerequisite for skipping* to it is *satisfied* if the tree $T(T_1 : T_2)$ contains a *branching* node labeled

$$(p, \omega_1) \equiv h(\bar{p}, \omega_2), \quad \text{where } \omega_1 \in \Gamma_1^{+} \text{ and } \omega_2 \in \Gamma_2^{+} \tag{6}$$

such that $N(p, \omega_1) \neq \emptyset$ with $|\omega_1| \geq |Q_1|$, and $N(\bar{p}, \omega_2) \neq \emptyset$ with $|\omega_2| \geq |Q_2|$.

Definition 12. Suppose that the prerequisite for skipping to the node labeled eq.(5) in question is satisfied as in Definition 11, and that

$$\langle (p, \omega_1) \equiv h(\bar{p}, \omega_2) \rangle \xrightarrow[T(T_1 : T_2)]{u \backslash x/v} \langle (q, \gamma) \equiv (\bar{q}, \bar{\gamma}) \rangle$$

and $\text{FIRST}_{live}(q, \gamma) = \text{FIRST}_{live}(\bar{q}, \bar{\gamma}) = \{\varepsilon\}$, where $\gamma = \varepsilon$ or $N(q, \gamma) = \emptyset$ with $1 \leq |\gamma| \leq |Q_1|$, and $\bar{\gamma} = \varepsilon$ or $N(\bar{q}, \bar{\gamma}) = \emptyset$ with $1 \leq |\bar{\gamma}| \leq |Q_2|$, for some $x \in (\Sigma \cup \{\lambda\})^*$, $u, v \in \Delta^*$ with $u = hv$, $q \in F_1$, $\bar{q} \in F_2$.

Now find a *shortest* string $\sigma(x_0) \in \Sigma^*$ such that

$$\langle (p, \omega_1) \equiv h(\bar{p}, \omega_2) \rangle \xrightarrow[T(T_1 : T_2)]{u_0 \backslash x_0/v_0} \langle (q, \gamma) \equiv (\bar{q}, \bar{\gamma}) \rangle,$$

$$(p, \omega_1) \xrightarrow[T_1]{\sigma(x_0)/u_0} (q, \zeta) \quad \text{and} \quad (\bar{p}, \omega_2) \xrightarrow[T_2]{\sigma(x_0)/v_0} (\bar{q}, \bar{\zeta})$$

with $|\zeta| \geq |\gamma|$ and $|\bar{\zeta}| \geq |\bar{\gamma}|$, for some $u_0, v_0 \in \Delta^*$, and check whether it is successful or not to have $u_0 = hv_0$. Then the skipping to the node in question is said to be *applicable* if the above checking is successful for every possible $(q, \gamma) \equiv (\bar{q}, \bar{\gamma})$ as above. A node labeled $(q, \gamma\alpha'') \equiv (\bar{q}, \bar{\gamma}\beta'')$ is defined to be a *skipping-end* from the node in question, and an edge label between them is defined to be $u_0 \backslash x_0/v_0$.

When skipping is applicable to the node in question, we expand it to have skipping-ends in *unchecked* status, then we turn the node in question to be *skipping*. The step of developing the comparison tree in this way is named *skipping* to the node labeled eq.(5) with respect to eq.(6). Whenever a new node is added to the comparison tree, we must apply skipping again to the node which has been already applied skipping, because it is possible that some new skipping-ends will be added to it afterward.

When the skipping has been applied to the node in question, the number of skipping-ends of it is at most $|F_1|(|Q_1| + 1) \times |F_2|(|Q_2| + 1)$.

4.4 The Whole Algorithm

The whole algorithm is shown in Fig. 1. Here, the next node to be visited is chosen as the "smallest" of the *unchecked* or *skipping* nodes, where the *size* of a node labeled $(p, \alpha) \equiv h(\bar{p}, \beta)$ is the pair $(\text{Max}\{|\alpha|, |\beta|\}, \text{Min}\{|\alpha|, |\beta|\})$, under lexicographic ordering.

Example 1. Let us apply our algorithm to the following pair of droct's:

$$T_1 = (\{p_0, p_1, p_2, p_3\}, \{A\}, \{a, b, c\}, \{a, b\}, \mu_1, p_0, \{p_3\}) \text{ and}$$

The Equivalence Checking Algorithm
Check whether $N(p, \alpha) \neq \emptyset$ or not for each configuration $(p, \alpha) \in Q_1 \times \Gamma_1^+$ with $1 \leq |\alpha| \leq |Q_1|$. Similarly, check whether $N(\bar{p}, \beta) \neq \emptyset$ or not for each configuration $(\bar{p}, \beta) \in Q_2 \times \Gamma_2^+$ with $1 \leq |\beta| \leq |Q_2|$.
Let the comparison tree consist of only a root labeled $(q_{01}, Z_{01}) \equiv (q_{02}, Z_{02})$ in *unchecked* status.
while the comparison tree contains an *unchecked* or a *skipping* node **do**
    if the comparison tree contains an *unchecked* node **then**
        let P be the smallest *unchecked* node
    else let P be the smallest *skipping* node **fi**
    suppose P is labeled $(p, \alpha) \equiv h(\bar{p}, \beta)$;
    if stack reduction is applicable to P **then**
        apply the stack reduction to P;
        turn the status of P to be *checked*, while its newly added son is in *unchecked*;
        turn the status of all *s-checked* nodes to be *skipping*
    else
        if $\mathrm{FIRST}_{live}(p, \alpha) = \mathrm{FIRST}_{live}(\bar{p}, \beta) = \{\varepsilon\}$ and both (p, α) and $(\bar{p}, \beta)$ are
          not in ε-mode **then**
          if $h = \varepsilon$ **then** turn P to *checked*
          else conclude that "$T_1 \not\equiv T_2$"; **halt fi**
        else if $(p, \alpha') \equiv h'(\bar{p}, \beta')$ $(h' \neq h)$ appears as the label of another internal
            node **then**
          conclude that "$T_1 \not\equiv T_2$"; **halt**
        else
          if $(p, \alpha) \equiv h(\bar{p}, \beta)$ appears as the label of another internal node **then**
            turn P to *checked*
          else
            if the prerequisite for skipping to P is satisfied **then**
                if skipping to P is applicable **then**
                    apply the skipping to P;
                    if either any skipping-end has been newly added to P as its son,
                      or any label of an edge from P has been changed by
                      the above skipping **then**
                      turn the status of P to be *skipping*, while its newly added sons
                      are in *unchecked*;
                      turn the status of all *s-checked* nodes to be *skipping*
                    else turn the status of P to be *s-checked* **fi**
                else conclude that "$T_1 \not\equiv T_2$"; **halt fi**
            else if output branch checking is successful for P **then**
                apply the branching to P;
                turn the status of P to be *checked*, while its newly added sons
                are in *unchecked*;
                turn the status of all *s-checked* nodes to be *skipping*
                else conclude that "$T_1 \not\equiv T_2$"; **halt fi fi fi fi fi fi od**
Conclude that "$T_1 \equiv T_2$"; **halt**

Fig. 1 The Equivalence Checking Algorithm

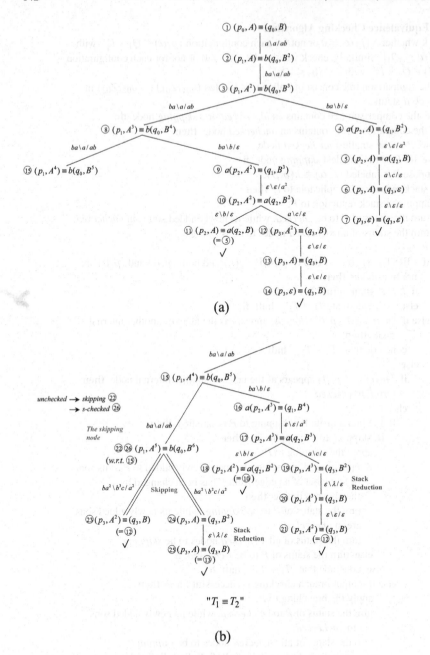

Fig. 2 The comparison tree for Example 1

$$T_2 = (\{q_0, q_1, q_2, q_3\}, \{B\}, \{a, b, c\}, \{a, b\}, \mu_2, q_0, \{q_3\}), \text{ where}$$

$$\mu_1 = \{(p_0, A) \xrightarrow{a/a} (p_1, A), (p_1, A) \xrightarrow{a/ba} (p_1, A^2), (p_1, A) \xrightarrow{b/ba} (p_2, \varepsilon),$$

$$(p_2, A) \xrightarrow{b/\varepsilon} (p_2, \varepsilon), (p_2, A) \xrightarrow{c/a} (p_3, A), (p_3, A) \xrightarrow{\varepsilon/\varepsilon} (p_3, \varepsilon)\} \text{ and}$$

$$\mu_2 = \{(q_0, B) \xrightarrow{a/ab} (q_0, B^2), (q_0, B) \xrightarrow{b/\varepsilon} (q_1, \varepsilon), (q_1, B) \xrightarrow{\varepsilon/a^2} (q_2, \varepsilon),$$

$$(q_2, B) \xrightarrow{b/\varepsilon} (q_2, \varepsilon), (q_2, B) \xrightarrow{c/\varepsilon} (q_3, \varepsilon)\}.$$

Successive application of branching steps yields an intermediate tree containing early nodes numbered ① – ⑲ in Fig. 2 (a) and (b). When ⑲$(p_3, A^3) \equiv (q_3, B^2)$ in Fig. 2 (b) is visited, stack reduction is applied to yield its son ⑳$(p_3, A^3) \equiv (q_3, B)$. And then, when ㉒$(p_1, A^4 \cdot A) \equiv b(q_0, B^5 \cdot B)$ is visited first, skipping is applied with respect to ⑮$(p_1, A^4) \equiv b(q_0, B^5)$ to yield skipping-ends ㉓$(p_3, A \cdot A) \equiv (q_3, \varepsilon \cdot B)$ and ㉔$(p_3, \varepsilon \cdot A) \equiv (q_3, B \cdot B)$. When the same node ㉖$(p_1, A^4 \cdot A) \equiv b(q_0, B^5 \cdot B)$ is visited again afterwards, the tree is not changed any more by this step. Then the algorithm halts with the correct conclusion that "$T_1 \equiv T_2$". (In fact, $\text{TRANS}(T_1) = \text{TRANS}(T_2) = \{a^i b^j c/(ab)^i a^2 \mid i > j \geq 1\}$.)

By using the proof similar to Ref. [10], we have the following theorem.

Theorem 1. *For the given two droct's which accept by final state, our equivalence checking algorithm halts in a polynomial-time with respect to the cardinalities of states, input symbols, output symbols, and transition-output rules with the correct conclusion.*

5 Conclusions

By extending the technique in Ref. [10], we have presented an algorithm for checking the equivalence of non-real-time droct's (i.e. droct's with possible ε-moves) which accept by final state. The worst-case time complexity of our algorithm is polynomial with respect to the description length of these droct's.

Acknowledgements. This work was supported in part by Grants-in-Aid for Scientific Research Nos.20500007 and 23500011 from the MEXT of Japan.

References

1. Higuchi, K., Tomita, E., Wakatsuki, M.: A polynomial-time algorithm for checking the inclusion for strict deterministic restricted one-counter automata. IEICE Trans. Inf. & Syst. E78-D(4), 305–313 (1995)
2. Higuchi, K., Wakatsuki, M., Tomita, E.: A polynomial-time algorithm for checking the inclusion for real-time deterministic restricted one-counter automata which accept by final state. IEICE Trans. Inf. & Syst. E78-D(8), 939–950 (1995)
3. Sénizergues, G.: L(A)=L(B)? decidability results from complete formal systems. Theoret. Comput. Sci. 251(1-2), 1–166 (2001)

4. Tajima, Y., Tomita, E., Wakatsuki, M., Terada, M.: Polynomial time learning of simple deterministic languages via queries and a representative sample. Theoret. Comput. Sci. 329, 203–221 (2004)
5. Tomita, E.: A direct branching algorithm for checking equivalence of some classes of deterministic pushdown automata. Inform. Control 52, 187–238 (1982)
6. Tomita, E., Seino, K.: A direct branching algorithm for checking the equivalence of two deterministic pushdown transducers, one of which is real-time strict. Theoret. Comput. Sci. 64, 39–53 (1989)
7. Tomita, E., Seino, K.: The extended equivalence problem for a class of non-real-time deterministic pushdown automata. Acta Informatica 32, 395–413 (1995)
8. Valiant, L.G.: Decision Procedures for Families of Deterministic Pushdown Automata. Ph.D. Thesis., Report No.7, University of Warwick Computer Centre (1973)
9. Valiant, L.G., Paterson, M.S.: Deterministic one-counter automata. J. Comput. System Sci. 10, 340–350 (1975)
10. Wakatsuki, M., Tomita, E., Nishino, T.: A polynomial-time algorithm for checking the equivalence for real-time deterministic restricted one-counter transducers which accept by final state. In: Proceedings of the 14th IEEE/ACIS International Conference on SNPD 2013, pp. 459–465 (2013)

An Unsupervised Ensemble Approach for Emotional Scene Detection from Lifelog Videos

Hiroki Nomiya, Atsushi Morikuni, and Teruhisa Hochin

Abstract. An emotional scene detection method is proposed in order to retrieve impressive scenes from lifelog videos. The proposed method is based on facial expression recognition considering that a wide variety of facial expression could be observed in impressive scenes. Most of conventional facial expression techniques adopt supervised learning methods. This is a crucial problem because preparing sufficient training data requires considerable human effort due to the diversity of facial expressions observed in lifelog videos. We thus propose a more efficient emotional scene detection method using an unsupervised facial expression recognition on the basis of cluster ensembles. Our approach does not require any training data sets and is able to detect various emotional scenes. The detection performance of the proposed method is evaluated through an emotional scene detection experiment.

Keywords: Lifelog, video retrieval, facial expression recognition, clustering, ensemble learning.

1 Introduction

Lifelog means a person's record of life and has recently attracted attention [1][2]. It can be recorded as various types of data such as texts, images, and videos. We focus on lifelog videos [3] because anyone can easily record his/her own lifelog videos due to the improved performance and the minimization of recently developed video cameras. Although lifelog videos will play an important role in recording and remembering one's life, they have a serious problem that it is difficult to accurately and efficiently retrieve useful scenes from the enormous amount of video data. As a result, valuable lifelog videos often remain unused.

Hiroki Nomiya · Atsushi Morikuni · Teruhisa Hochin
Department of Information Science, Kyoto Institute of Technology,
Goshokaido-cho, Matsugasaki, Sakyo-ku, Kyoto, 606-8585, Japan
e-mail: {nomiya,hochin}@kit.ac.jp,m2622043@edu.kit.ac.jp

© Springer International Publishing Switzerland 2015   145
R. Lee (ed.), *SNPD*,
Studies in Computational Intelligence 569, DOI: 10.1007/978-3-319-10389-1_11

In order to solve the issue, we propose an impressive scene detection method for lifelog video retrieval. Impressive scenes are considered to be useful since they contain recordings of important events that need to be retrieved during the lifelog video retrieval. We attempt to obtain impressive scenes by detecting emotional scenes based on the fact that it is important for human communication to utilize nonverbal communications including emotional expressions [4]. We focus on facial expression recognition for the detection of emotional scenes because most of emotions can be reflected in the facial expressions.

Facial expression recognition has been extensively studied and can be applied to video-scene detection [5][6][7]. However, most of facial expression recognition techniques are based on supervised learning. They generally require a large amount of training data to construct a good facial expression recognition model. Since it is quite difficult to automatically collect training data, preparing sufficient training data requires considerable human effort.

In our approach, we aim to improve the efficiency of the facial expression recognition by introducing an unsupervised learning framework using a clustering technique. The facial expression is recognized by classifying the facial image into one of the clusters corresponding to a certain facial expression.

For the purpose of improving the recognition accuracy, we introduce an ensemble learning approach called cluster ensemble [8]. The cluster ensemble can enhance and stabilize the accuracy of facial expression recognition by combining diverse sets of clusters into a single set of more discriminative clusters.

In order to detect emotional scenes, we introduce an efficient emotional scene detection method on the basis of the result of the facial expression recognition for each frame image in a video. Because the ensemble clustering is unsupervised, the proposed method neither requires learning data nor the predefinition of facial expressions. We show the effectiveness of the the proposed method through an emotional scene detection experiment.

The remainder of this paper is organized as follows. Section 2 presents some related works. Section 3 describes the facial features used to recognize facial expressions. Section 4 elaborates the facial expression recognition method using the cluster ensemble. Section 5 explains the emotional scene detection method. Section 6 shows the emotional scene detection experiment using several lifelog videos. Finally, Section 7 concludes this study.

2   Related Works

Facial expression recognition plays the most important role in our emotional scene retrieval. In order to precisely and efficiently recognize facial expressions, various kinds of facial expression recognition techniques have been proposed. The performance of the facial expression recognition can be greatly influenced by the facial features. Currently, there are two major types of facial features, appearance features and geometric features [9].

The appearance features are based on the skin texture of a face and can describe the appearance changes of a face such as wrinkles and furrows. The appearance features can be obtained from the intensity distributions of the pixels in a facial image. For instance, Gabor wavelet [10] and the local binary patterns [11] are widely used as this type of features.

The geometric features describe the shape and locations of several facial components such as eyebrows, eyes, and a mouth. For example, 3D models of the faces are used to accurately describe and recognize facial expressions [12][13]. These models can properly describe the facial structures and will be effective for accurate facial expression recognition. In the lifelog video retrieval, however, it will be difficult to prepare 3D facial features within reasonable cost. By using several salient facial feature points (e.g., the end points of the mouth and a center point of the eyes), the facial features can be more concise. The facial features are defined as the positional relationship of the facial feature points such as the distance between two points and the angle between two line segments formed by connecting three points [14][15][16]. In this study, we adopt the geometric features represented by the positional relationships of a few facial feature points because of the conciseness and the better understandability of the facial features.

Most of the facial expression recognition methods are supervised while the supervised learning needs sufficient training data. Because preparing the training data requires considerable human resource, it is desirable to construct the facial expression models in an unsupervised manner. There exist the unsupervised facial expression methods on the basis of unsupervised machine learning techniques using such as principal component analysis [17][18]. Considering that lifelog video databases can be very large, the facial expression recognition process should be fully efficient. Although the efficient facial expression recognition and emotional scene detection methods are proposed [19], the accuracy is not adequate. In this study, we aim to develop an unsupervised emotional scene detection method considering both accuracy and efficiency.

3 Facial Features

Prior to the emotional scene detection, the facial expression recognition is performed for each frame image in a video. In order to discriminate the facial expressions, we define several facial features on the basis of the positional relationships of several salient points on the face (we call them *facial feature points*).

3.1 Facial Feature Points

We utilize a total of 59 facial feature points. They are located on the eyebrows (10 points), eyes (22 points), a nose (9 points), a mouth (14 points), and nasolabial folds (4 points) as shown in Fig.1. The facial feature points are obtained by using a software application called FaceSDK 4.0 [20]. The facial feature points are denoted by $p_1, \ldots, p_{59}$.

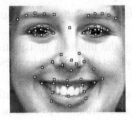

Fig. 1 Facial feature points

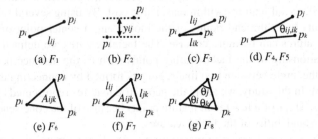

Fig. 2 Facial features

3.2 Facial Feature Values

Some kind of facial features such as the ones defined in the Facial Action Coding System [21] are considered to be useful for the recognition of facial expressions. However, there is the personal differences in regard to the appearance and intensity of facial expressions. Hence, it will be effective to personalize the facial feature values used to recognize facial expressions.

We thus define eight types of facial features by combining two or three facial feature points taking the computational efficiency into account. There are a large number of possible combinations (i.e., possible feature values) on the basis of this definition. Several useful facial feature values can be selected for each person to recognize his/her facial expression accurately.

- **Length of a line segment (F_1)**

This facial feature value is the normalized length of the line segment l_{ij} formed by connecting two facial feature points p_i and p_j as shown in Fig. 2 (a). The length is normalized by dividing it by the Euclidean distance d between the center points of the left and right eyes in order to make the feature value independent of the scale of the facial image. This facial feature can represent the relative positional relationship between two facial components (e.g., an eyebrow and an eye).

- **Absolute value of the difference of the y-coordinate (F_2)**

This is the normalized absolute value of the difference of the y-coordinate y_{ij} between p_i and p_j as shown in Fig. 2 (b). The absolute difference is divided by d for the normalization. This facial feature is defined taking into consideration the tendency that the y-coordinates of the facial feature points can be largely change when a certain kind of facial expression (e.g., surprise or laughter) appears.

- **Ratio of the lengths of two line segments (F_3)**

This is the ratio of the length of the line segment l_{ij} to that of the line segment l_{ik} as shown in Fig. 2 (c). These line segments are formed by three facial feature points p_i, p_j, and p_k. This facial feature can represent the relative positional relationship between three facial components.

- **Angle between two line segments (F_4)**

This is the angle $\theta_{ij,ik}(0 \leq \theta_{ij,ik} \leq \pi)$ between the line segments l_{ij} and l_{ik}. This facial feature is shown in Fig. 2 (d). We attempt to make use of the shape of a facial component using this feature value. For example, the shape of a corner of a mouth is associated with this feature value obtained from the facial feature points near the corner of a mouth.

- **Cosine of the angle between two line segments (F_5)**

This facial feature value is defined as the cosine of F_4, that is, $\cos\theta_{ij,ik}$. It represents the shape of a facial component from different aspect.

- **Area of a triangle (F_6)**

This is the normalized area of the triangle A_{ijk} formed by three facial feature points p_i, p_j, and p_k as shown in Fig. 2 (e). The area is divided by d^2 for the normalization. This facial feature can be an approximation of the area of a facial component.

- **Ratio of the area of a triangle to the squared sum of the lengths of three sides of the triangle (F_7)**

This is the ratio of the area of the triangle A_{ijk} to the square of the sum of the lengths of the three sides l_{ij}, l_{ik}, and l_{jk}. This facial feature is shown in Fig. 2 (f) and its value is $A_{ijk}/(l_{ij} + l_{ik} + l_{jk})^2$. It is defined based on the shape of a triangle as well as the area of a triangle.

- **Internal angles of a triangle (F_8)**

This feature value is based on the three internal angles $\theta_{ij,jk}$, $\theta_{ij,ik}$, and $\theta_{ik,jk}$ in the triangle shown in Fig. 2 (g). It is computed by the equation $|\theta_{ij,jk} - \frac{\pi}{3}| + |\theta_{ij,ik} - \frac{\pi}{3}| + |\theta_{ik,jk} - \frac{\pi}{3}|$. This feature value is associated with the shape of a triangle and it is minimized (becomes zero) when the triangle is a regular triangle.

These facial feature values are defined for all the possible combinations of the facial feature points. The number of F_1 or F_2 is 1711 ($= {}_{59}C_2$), that of F_3, F_4, or F_5 is

$97527 (= 3 \times {}_{59}C_3)$, and that of F_6, F_7, or F_8 is $32509 (= {}_{59}C_3)$. Therefore, the total number of facial feature values is 393530.

The facial features belonging to F_i are denoted by $\{f_{i,1}, \ldots, f_{i,n_i}\}$, where n_i is the number of facial features belonging to F_i. That is, $n_1 = n_2 = 1711$, $n_3 = n_4 = n_5 = 97527$, and $n_6 = n_7 = n_8 = 32509$. For instance, $f_{1,1}$ is the normalized length of the line segment formed by p_1 and p_2. Similarly, $f_{1,2}$ is computed from p_1 and p_3.

4 Facial Expression Recognition

The proposed facial expression recognition is performed by clustering the frame images in a video. In our clustering approach, a single cluster represents a single facial expression. The clustering is done for each type of facial feature (i.e., F_1 to F_8) on the basis of k-means method. Then, the resulting eight sets of clusters are integrated into a single set of clusters by the ensemble clustering.

4.1 Ensemble Clustering

Ensemble clustering is a kind of metaclustering method that integrates diverse sets of clusters (called *weak clusters*) into a single set of clusters (called *strong clusters*). Generally, the discrimination ability and the robustness to outliers of the strong clusters are better than those of weak clusters.

There are several clustering methods to generate weak clusters. In the proposed method, we use k-means method due to its conciseness. There also exist several ensemble clustering method. We use the cluster-based similarity partitioning algorithm (CSPA) because it has relatively small computational complexity [8].

4.2 Constructing Weak Clusters

A total of eight sets of weak clusters are constructed using the facial features F_1 to F_8. A set of weak clusters for F_i is represented as $\{C_{i,1}, \ldots, C_{i,K}\}$ $(i = 1, \ldots, 8)$, where K is the number of clusters and is equivalent to the number of facial expressions. This set of weak clusters is generated based on k-means method by using a set of feature vectors $\{X_{i,1}, \ldots, X_{i,N}\}$ as its input. N is the number of frame images. $X_{i,j}$ is a feature vector obtained from the j-th frame image and can be given by the equation $X_{i,j} = (f_{i,1}(x_j), \ldots, f_{i,n_i}(x_j))$. Here, $f_{i,k}(x_j)$ is the feature value of $f_{i,k}$ for the j-th frame image.

4.3 Feature Selection

Using all the feature vectors to construct a set of weak clusters imposes considerable computational cost because the number of facial feature values (n_i) is quite large. In order to reduce the computational cost, we define the usefulness of the facial features and make use of a small number of useful facial feature values. The usefulness is defined on the basis of the Gaussian mixture model and the ratio of the between-class variance to the within-class one of the feature values.

The weak clusters are constructed by using only a few facial feature values with high usefulness values. That is, the feature vectors $\{X'_{i,1},\ldots,X'_{i,N}\}$ are used as the input of the k-means clustering algorithm. $X'_{i,j}$ is given as $X'_{i,j}=(f_{i,r_1}(x_j),\ldots,f_{i,r_m}(x_j))$, where f_{i,r_k} is the k-th most useful facial feature value (i.e., the facial feature value with the k-th highest usefulness value). The number of feature values m is determined experimentally. The algorithm to determine the usefulness of a facial feature value $f_{i,j}$ is shown in Algorithm 1.

Algorithm 1. Determination of the usefulness of a facial feature value

---

Input: A set of facial feature values for each frame image $\phi = \{f_{i,j}(x_1),\ldots,f_{i,j}(x_N)\}$ and the number of weak clusters K.

Procedure:

1: Compute the Gaussian mixture model P by using ϕ through the EM algorithm. The number of Gaussian distributions is K. P is given by Equation (1).

$$P(f_{i,j}(x)) = \sum_{k=1}^{K} w_k \frac{1}{\sqrt{2\pi\sigma_k^2}} \exp\left(-\frac{(f_{i,j}(x)-\mu_k)^2}{2\sigma_k^2}\right) \tag{1}$$

where, w_k is the mixing coefficient of the k-th Gaussian distribution. μ_k and σ_k^2 denote the mean and variance of the k-th distribution, respectively.

2: Compute the usefulness value U using Equation (2).

$$U = \frac{\sum_{k=1}^{K}(\mu_k-\mu)^2}{\sum_{k=1}^{K}\sigma_k^2} \tag{2}$$

where, μ is the mean value of $\mu_k(k=1,\ldots,K)$. Note that we do not use the mixture coefficients in the computation of the usefulness values.

4.4 Construction of Strong Clusters

The eight sets of weak clusters are integrated into a set of strong clusters according to the CSPA-based ensemble clustering algorithm.

First, a similarity matrix is computed from the sets of weak clusters. An element in the similarity matrix corresponds to the similarity between the facial feature values of arbitrary two frame images. The similarity is defined as the number of weak clusters containing both of the two frame images. In this case, the minimum and maximum values of the similarity are 0 and 8 respectively, because there are eight sets of weak clusters.

Next, the initial strong clusters are defined so that a strong cluster contains a single frame image. Then, two most similar strong clusters are integrated into a single strong cluster. The similarity is determined on the basis of the similarity matrix. The integration step is repeated until the number of strong clusters becomes K.

The resulting K strong clusters can discriminate K types of facial expressions. The algorithm to construct strong clusters is shown in Algorithm 2.

Algorithm 2. Construction of the strong clusters

Input: The number of strong clusters K (note that this is the same as the number of weak clusters in a set of weak clusters) and the weak cluster sets $\{C_{1,1},\ldots,C_{1,K}\},\ldots,\{C_{8,1},\ldots,C_{8,K}\}$.

Procedure:

1: Compute the similarity matrix S from Equation (3).

$$S = \begin{pmatrix} s_{1,1} & s_{1,2} & \cdots & s_{1,N} \\ s_{2,1} & s_{2,2} & \cdots & s_{2,N} \\ \vdots & \vdots & \ddots & \vdots \\ s_{N,1} & s_{N,2} & \cdots & s_{N,N} \end{pmatrix} \tag{3}$$

where $s_{i,j}$ is the number of weak clusters that contain both frame images x_i and x_j. Therefore, $0 \le s_{i,j} \le 8$. S is symmetric because $s_{i,j} = s_{j,i}$.

2: Generate N initial sets of strong clusters $G_1,\ldots,G_N$ such that $G_i = \{x_i\}$.

3: $v \leftarrow N$.

4: Integrate G_{j^*} into G_{i^*} ($G_{i^*} \leftarrow G_{i^*} \cup G_{j^*}$). The subscripts i^* and j^* ($i^* < j^*$) are determined so that Γ in Equation (4) is maximized.

$$\Gamma = \frac{1}{|G_{i^*}||G_{j^*}|} \sum_{a \in G_{i^*}} \sum_{b \in G_{j^*}} s_{ab} \tag{4}$$

5: Renumber the subscripts of the strong cluster sets to be $G_1,\ldots,G_{v-1}$.

6: $v \leftarrow v - 1$.

7: If $v > K$ then return to step 4. Otherwise, finish the clustering and output the strong clusters $G_1,\ldots,G_K$. Note that our method judge that the same facial expression appears in the frame images x_i and x_j if they are classified into the same strong cluster.

---

5 Emotional Scene Detection

The emotional scenes are detected from a video according to the result of the facial expression recognition for each frame image. An emotional scene with a certain facial expression is determined by using the frame images in the corresponding strong cluster.

At the first step of the emotional scene detection, each frame image in the strong cluster is regarded as a single emotional scene. Then, neighboring emotional scenes are integrated into a single emotional scene. The integration process is repeated until no more emotional scenes can be integrated.

The resulting emotional scenes are regarded as the emotional scenes for the strong cluster. As a result, a total of K sets of emotional scenes can be detected from K strong clusters. The algorithm to detect the emotional scenes is shown in Algorithm 3.

Algorithm 3. Emotional scene detection

Notations:

   E_i^c: The i-th emotional scene in which the facial expression c appears.

   $first(E_i^c)$: The frame number of the beginning frame in E_i^c.

   $last(E_i^c)$: The frame number of the ending frame in E_i^c.

   $length(E_i^c)$: The length of E_i^c. It is equivalent to $last(E_i^c) - first(E_i^c) - 1$.

   $\#int(E_i^c)$: The number of emotional scenes integrated into E_i^c.

   $\#nonemo(E_i^c)$: The number of nonemotional frames in E_i^c. Note that a nonemotional frame means that the facial expression appears in that frame is different from c.

   $dist(E_i^c, E_j^c)$: The distance between E_i^c and E_j^c $(i < j)$. It is equivalent to $first(E_j^c) - last(E_i^c) - 1$.

Initialize: For each frame image classified into the facial expression c, initialize according to Equation (5):

$$first(E_i^c) = last(E_i^c) = c_i, \ \#int(E_i^c) = 0,$$

$$\#nonemo(E_i^c) = 0, \ length(E_i^c) = 1, \ (1 \leq i \leq M_c) \tag{5}$$

where, c_i is the frame number of the i-th emotional frame in the video. M_c is the number of emotional frames. An emotional frame is the frame classified into c. That is, each emotional scene consists of a single emotional frame.

Procedure:

1: Find i^* in accordance with Equation (6):

$$i^* = \underset{i}{\operatorname{argmin}} \ dist(E_i^c, E_{i+1}^c)$$

$$s.t. \ dist(E_i^c, E_{i+1}^c) \leq \frac{length(E_i^c) - \#nonemo(E_i^c)}{\#int(E_i^c) + 1}$$

$$\wedge \ dist(E_i^c, E_{i+1}^c) \leq \frac{length(E_{i+1}^c) - \#nonemo(E_{i+1}^c)}{\#int(E_{i+1}^c) + 1} \tag{6}$$

2: If there is no i^* that satisfies Equation (6), finish the procedure and output current emotional scenes. Otherwise, proceed to step 3.

3: Integrate $E_{i^*+1}^c$ into $E_{i^*}^c$ by updating $E_{i^*}^c$ as follows:

$$last(E_{i^*}^c) \leftarrow last(E_{i^*+1}^c), \#int(E_{i^*}^c) \leftarrow \#int(E_{i^*}^c) + 1,$$

$$\#nonemo(E_{i^*}^c) \leftarrow \#nonemo(E_{i^*}^c) + first(E_{i^*+1}^c) - last(E_{i^*}^c) - 1$$

Note that $length(E_{i^*}^c)$ is also updated due to the update of $last(E_{i^*}^c)$.

4: Delete $E_{i^*+1}^c$ and renumber the subscripts of E_i^c so that the emotional scenes become $E_1^c, \ldots, E_{M_c-1}^c$.

5: $M_c \leftarrow M_c - 1$ and return to step 1.

---

6 Experiment

6.1 Experimental Settings

We prepared five lifelog videos by five subjects termed Subject A, B, C, D, and E. All the subjects are male university students.

The lifelog videos contain the scenes of playing cards recorded by web cameras. A single web camera recorded a single subject so that the subject's frontal face was recorded. This experimental setting is due to the limitation of FaceSDK that it can detect the facial feature points of a single frontal face. The length of each video is approximately five minutes. The size of each video is 640×480 pixels and the frame rate is 30 frames per second. Considering the high frame rate, we selected frames from each video after every 10 frames in order to reduce the computational cost. Consequently, we used 900 frame images for each video.

Apart from the web cameras, a video camera was used to record the scenes of playing cards including all subjects at the same time. The videos were recorded in order to show the results of the emotional scene retrieval to the users while they were not used in this experiment. Note that the videos recorded are not shown in this paper because of privacy reasons.

The facial expressions observed in most of the emotional scenes in the lifelog videos were smiles. Thus, we set the value of K to 2 intending to detect the emotional scenes with smiles, that is, to discriminate smiles and other facial expressions (mainly neutral faces).

6.2 Emotional Scene Detection Accuracy

The F-measure of the emotional scene detection is computed for the evaluation of the performance of the proposed method. The F-measure is defined by Equation (7) using the recall and precision defined by Equation (8) and (9), respectively.

$$F-measure = \frac{2 \cdot recall \cdot precision}{recall + precision} \tag{7}$$

$$recall = \frac{|T \cap \hat{T}|}{|T|} \tag{8}$$

$$precision = \frac{|T \cap \hat{T}|}{|\hat{T}|} \tag{9}$$

where, T is the correct set of emotional frames. One of the authors determined whether each frame was emotional prior to the experiment. $\hat{T}$ is the set of emotional frames detected by the proposed method.

We set the number of facial feature values m selected in the feature selection from 1 to 20. The F-measure of the emotional scene detection for each subject is shown in Fig. 3. Note that these are the average values of five trials for each subject using different seeds for the randomization in the k-means algorithm.

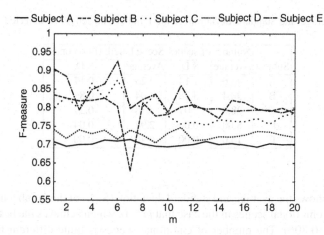

Fig. 3 F-measure of the emotional scene detection for each subject

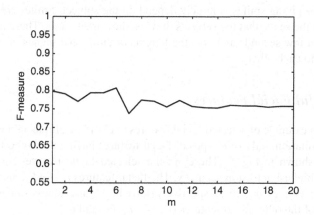

Fig. 4 Average F-measure of five subjects

The value of F-measure varies from subject to subject. This will be caused by the difference of the intensity of the smiles. Some subjects often give little smiles while others tend to give full smiles. The former will make the detection of emotional scenes more difficult, the latter will make it relatively easy. Therefore, the accuracy of emotional scene detection depends to some extent on the person.

The number of feature values m has an influence on the detection accuracy. On average, the F-measure slightly decreases in proportion to the number of feature values when $m > 6$ as shown in Fig. 4. This means a small number of feature values are sufficient for the emotional scene detection. Using large number of feature values will not be effective because it increases the number of redundant feature values and decreases the average usefulness of the feature values.

Table 1 Average and standard deviation of the number and length of detected emotional scenes

Subject	Number of scenes		Scene length (in seconds)	
	Average	S.D.	Average	S.D.
A	46.6	1.14	2.25	0.17
B	30.4	3.78	4.04	0.83
C	28.4	2.07	4.38	0.27
D	8.00	0.00	2.97	0.04
E	15.0	0.00	6.40	0.00

Table 1 shows the average and standard deviation (S.D.) of the number and length of detected emotional scenes in the case that $m = 6$, which achieves the best average F-measure (0.806). The number of emotional scenes is quite different from subject to subject, while the average lengths of the scenes are relatively similar to each other. The difference of the number of emotional scenes stems from the fact that the frequency of giving smiles is largely depend on the subject. Smiles are frequently observed in the scene that the subjects draw or show their cards. These actions generally take a few seconds and thus the lengths of emotional scenes of the subjects are similar to each other.

6.3 Useful Facial Features

We show an example of selected facial features to clarify useful facial features. Due to space limitation, only four types of facial features having the highest usefulness values are shown in Fig. 5. These are the selected facial features for Subject D having the highest detection accuracy. The facial features F_3, F_4, F_5, and F_8 (shown in Fig. 5 (a), (b), (c), and (d), respectively) have higher usefulness values compared with those of the other facial features (i.e., F_1, F_2, F_6, and F_7).

Most of selected facial features contain the facial feature points on the mouth and/or nasolabial folds. These points will be useful for the detection of smiles because their positions tend to saliently change as a smile appears. From this result, our feature selection method seems to work well.

6.4 Emotional Scene Detection Efficiency

For the evaluation of the efficiency of the proposed method, we show the average computation time on the five subjects and its S.D. for the emotional scene detection in Table 2. In this table, the emotional scene detection process is broken down into three steps: the feature value computation, the facial expression recognition, and the emotional scene detection.

A computer with a Xeon W3580 CPU (3.33GHz) and 8GB memory was used and parallel processing was not used in this experiment. Note that the processing time

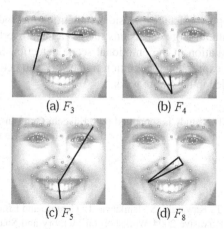

(a) F_3 (b) F_4

(c) F_5 (d) F_8

Fig. 5 Example of selected features for Subject D

Table 2 Average and standard deviation of the computation time (in seconds)

Process	Average	S.D.
Feature computation	0.787	0.045
Facial expression recognition	1.656	0.305
Emotional scene detection	0.001	0.000
Total	2.443	0.301

to obtain the facial feature points is not included because this process is performed by an existing software application.

Facial expression recognition requires slightly long computation time due to the ensemble clustering. On the other hand, the emotional scene detection is considerably fast owing to the concise detection algorithm. As a whole, the proposed method is fully efficient because it takes less than three seconds to detect emotional scenes from a 5-minute-long video.

7 Conclusion

This paper proposed an emotional scene detection method for the better utilization of lifelog videos. The proposed method is suitable for various lifelog videos because it is based on unsupervised learning and does not require any training data sets. In addition, the proposed method is fully efficient by introducing an ensemble learning framework using a few useful feature values.

The experimental results show the effectiveness of the proposed method to detect the emotional scenes with smiles. However, the performance of the proposed method for the other emotions is unclear. Thus, evaluating the proposed method for a wider variety of emotions is included in the future work.

Currently, the proposed facial expression recognition method can be applied for only frontal faces without large occlusions such as a hand overlapping a face. This can be a crucial limitation in the emotional scene detection from lifelog videos. Improving the facial expression recognition framework and getting rid of this limitation are also included in the future work.

Acknowledgements. This study is supported by a Kakenhi Grant-in-Aid for Young Scientists (B) (22700098) from the Japan Society for the Promotion of Science (JSPS).

References

1. Aizawa, K., Hori, T., Kawasaki, S., Ishikawa, T.: Capture and Efficient Retrieval of Life Log. In: Proc. of Pervasive 2004 Workshop on Memory and Sharing Experiences, pp. 15–20 (2004)
2. Gemmell, J., Bell, G., Luederand, R., Drucker, S., Wong, C.: MyLifeBits: Fulfilling the Memex Vision. In: Proc. of the 10th ACM International Conference on Multimedia, pp. 235–238 (2002)
3. Datchakorn, T., Yamasaki, T., Aizawa, K.: Practical Experience Recording and Indexing of Life Log Video. In: Proc. of the 2nd ACM Workshop on Continuous Archival and Retrieval of Personal Experiences, pp. 61–66 (2005)
4. Mehrabian, A.: Silent Messages. Wadsworth, Belmont (1971)
5. Datcu, D., Rothkrantz, L.: Facial Expression Recognition in Still Pictures and Videos Using Active Appearance Models: A Comparison Approach. In: Proc. of the 2007 International Conference on Computer Systems and Technologies, pp. 1–6 (2007)
6. Fanelli, G., Yao, A., Noel, P.-L., Gall, J., Van Gool, L.: Hough Forest-based Facial Expression Recognition from Video Sequences. In: Kutulakos, K.N. (ed.) ECCV 2010 Workshops, Part I. LNCS, vol. 6553, pp. 195–206. Springer, Heidelberg (2012)
7. Littlewort, G., Bartlett, M.S., Fasel, I., Susskind, J., Movellan, J.: Dynamics of Facial Expression Extracted Automatically from Video. Journal of Image and Vision Computing 24(6), 615–625 (2004)
8. Strehl, A., Ghosh, J.: Cluster Ensembles – a Knowledge Reuse Framework for Combining Multiple Partitions. Journal of Machine Learning Research 3, 583–617 (2003)
9. Tian, Y., Kanade, T., Cohn, J.F.: Facial Expression Recognition. In: Li, S.Z., Jain, A.K. (eds.) Handbook of Face Recognition, pp. 487–519. Springer, London (2011)
10. Lyons, M., Akamatsu, S.: Coding Facial Expressions with Gabor Wavelets. In: Proc. of the 3rd International Conference on Automatic Face and Gesture Recognition, pp. 200–205 (1998)
11. Feng, X., Pietikäinen, M.: Facial Expression Recognition with Local Binary Patterns and Linear Programming. Pattern Recognition and Image Analysis 15(2), 546–548 (2005)
12. Wang, J., Yin, L., Wei, X., Sun, Y.: 3D Facial Expression Recognition Based on Primitive Surface Feature Distribution. In: Proc. of IEEE Computer Society Conference on Computer Vision and Pattern Recognition, pp. 1399–1406 (2006)
13. Soyel, H., Demirel, H.O.: Facial Expression Recognition Using 3D Facial Feature Distances. In: Kamel, M.S., Campilho, A. (eds.) ICIAR 2007. LNCS, vol. 4633, pp. 831–838. Springer, Heidelberg (2007)
14. Esau, N., Wetzel, E., Kleinjohann, L., Kleinjohann, B.: Real-time Facial Expression Recognition Using a Fuzzy Emotion Model. In: Proc. of International Conference on Fuzzy Systems, pp. 1–6 (2007)

15. Hupont, I., Cerezo, E., Baldassarri, S.: Sensing Facial Emotion in a Continuous 2D Affective Space. In: Proc. of International Conference on Systems, Man, and Cybernetics, pp. 2045–2051 (2010)
16. Nomiya, H., Morikuni, A., Hochin, T.: Emotional Video Scene Detection from Lifelog Videos using Facial Feature Selection. In: Proc. of 4th International Conference on Applied Human Factors and Ergonomics, pp. 8500–8509 (2012)
17. Hoey, J.: Hierarchical Unsupervised Learning of Facial Expression Categories. In: Proc. of IEEE Workshop on Detection and Recognition of Events in Video, pp. 99–106 (2001)
18. Gholami, B., Haddad, W.M., Tannenbaum, A.R.: An Unsupervised Learning Approach for Facial Expression Recognition Using Semi-Definite Programming and Generalized Principal Component Analysis. In: Proc. of Image Processing: Algorithms and Systems, pp. 1–10 (2010)
19. Nomiya, H., Morikuni, A., Hochin, T.: Impressive Scene Detection from Lifelog Videos by Unsupervised Facial Expression Recognition. In: Proc. of 14th International Conference on Software Engineering, Artificial Intelligence, Networking and Parallel/Distributed Computing, pp. 444–449 (2013)
20. Luxand Inc., Luxand FaceSDK 4.0, http://www.luxand.com/facesdk (April 18, 2014)
21. Ekman, P., Friesen, W.: Unmasking the Face: A Guide to Recognizing Emotions from Facial Clues. Prentice Hall, Englewood Cliffs (1975)

A Discovery Method of Anteroposterior Correlation for Big Data Era

Takafumi Nakanishi

Abstract. In this paper, we present a new knowledge extraction method on Big data era. We introduce new concepts, anteroposterior correlation, and propose an extraction method of anteroposterior correlation. The anteroposterior correlation means the correlation based on the time anteroposterior relation. We consider that Heterogeneity, continuity, and visualization are the most critical features of Big data analytics, which provides a scale and connection merits based on them. No current data analysis methods are based on opened assumptions. Big data analytics provides a new data analysis method based on opening assumptions. In this paper, we especially focus on an aspect of heterogeneity. We discover a correlation in consideration of the continuity of time. By our method, we effectively discover relationships between heterogeneous things, events and phenomena. The anteroposterior correlations are represented in relative comparison with each conditional probability distribution. The one of the features of our method is a measurement correlation by using conditional probability. That is, we calculate the correlation relative by representing all in conditional probability, no absolutely. Our method is determined higher correlation by comparison to each heterogeneous thing, event and phenomenon. This is the most important points on the Big data era. When you apply current association rule extraction techniques, you obtain too big rule base to organize them. By our method, we realize the one of the methods for decision mining.

1 Introduction

We information science researchers have already constructed data sensing, aggregation, retrieval, analysis, and visualization environment by web portals, software,

Takafumi Nakanishi
The Center for Global Communications (GLOCOM)
International University of Japan, Tokyo, Japan
e-mail: takafumi@glocom.ac.jp

© Springer International Publishing Switzerland 2015
R. Lee (ed.), *SNPD*,
Studies in Computational Intelligence 569, DOI: 10.1007/978-3-319-10389-1_12

APIs, etc. on the web. The exponential growth in the amount of field data means that innovative measures are needed for data management, analysis, and accessibility. Therefore, almost users' needs are changing. Online databases have become crucial to access and publish various data depending on the user's purpose, task, or interest.

First, for leading the solutions, we have to marshal what Big Data is. We consider that the Big Data includes two prominent types of direction for ICT research purposes.

The one is the scale and the speed issue of data processing. A lot of researchers have done this theme such as HPC, parallel distributed processing, and etc. Another is a schemaless data processing issue. It is important to real-timely discover the answers or clues for a user. A system has to create an appropriate schema from the data themselves given by a user's query. Until now, data organized on the database schema. Currently, there are only fragmental various data on the web. This is a big paradigm shift. That is, it is necessary to create the schema and data structures corresponding to user's required processing after a user inputs some queries. We have to shift the system from closed assumption to opened assumption. In the case of the closed assumption, the schema was designed in advance in consideration of orthogonal or independence. In the case of the opened assumption, we cannot care orthogonal and independence when the system dynamically creates schema.

On this background, it is important to research new data analytics method. Especially, it is necessary to discover the relation between each thing, event and phenomenon. By this discovery, we can do risk aversion, can pursue a cause, or can predict the phenomenon which may happen in the near future. On the other viewpoint, we consider that the essence of Big data is not only massive data processing, but also optimization of the real world by the knowledge acquired from aggregated data. The current tendency of the research on Big data is how to aggregate massive data and how to process these data quickly. The future tendency of researches will become discovery methods of the optimized solution from the Big data. It is important to realize risk aversion, the cause unfolding, and the phenomenon prediction in the near future. In the first step, it is necessary to extract relationships between each heterogeneous thing, events and phenomena.

We propose a new concept, anteroposterior correlation. The anteroposterior correlation means the correlation based on the time anteroposterior relation. The anteroposterior correlations are represented in relative comparison with each conditional probability distribution. The one of the most important points is relatively comparable. In the other words, this system discovers the relation with higher correlation by comparing each value of a conditional probability as each weight. In this paper, we represent a new discover method of anteroposterior correlation between each heterogeneous thing, events and phenomena. We discover a correlation in consideration of the continuity of time. By our method, we effectively discover relationship between heterogeneous things, events and phenomena. The one of the features of our method is a measurement correlation by using conditional probability. We calculate the anteroposterior correlation relative by representing all in conditional probability.

By our method, we clarify the relationship between each heterogeneous thing, event and phenomenon.

The person harnesses the past experience and does prediction and prevention. Our system can do the same things like this. That is, we can predict and prevention of Big data. Therefore, we realize the one of the methods for decision mining.

The contributions of this paper are as follows:

- We summarize essences of Big data analytics.
- We propose a new concept for discovery, an anteroposterior correlation.
- We represent a discovery method of anteroposterior correlation between each heterogeneous thing, event and phenomenon.

This paper is organized as follows. In section 2, we survey the related works of our proposed method. In section 3, we introduce our ethic of essences for Big data analytics. After that, we propose our method, a discovery method of anteroposterior correlation in section 4. We present some experiment results and discussion in section 5. Finally, we conclude in section 6.

2 Related Works

Correlation or similarity measures for a discovery of relationships have been studied for a long time. The most popular and basic method is vector space model [1].

Dimensionality reduction techniques of vector space model have been used for developing traditional vector space models such as latent semantic indexing [2] and the mathematical model of meaning [3, 4]. These techniques are applied to information resources, characterized by elements in a flat domain. However, it is to be noted that when the elements have a hierarchical structure, all the elements are not orthogonal to each other. A few studies have used computational measures of directionality relationships [5] in an orthogonal vector space. The mathematical model of meaning realizes a context-driven dynamic semantic computation. However, it has to prepare a space for the semantic commutation before. Our method processes a dynamic data-driven space creation corresponding to a context. That is, our method does not have to prepare the space before. It is very an important difference, because we cannot create the space or any schemas before in the open assumption. Currently, we are in Big data era. In the Big data environment, we can aggregate various and a lot of fragmental data. We cannot predict what kinds of the data we obtain in advance. Actually, the rise of the key-value store means that the schema cannot be designed in advance. Since the data updates are faster and faster, we should change dynamically also space for the semantic computations and analysis.

There have been studies defining similarity metrics for a discovery of relationships, such as WordNet [6]. Rada et al. [7] have proposed a "conceptual distance" that indicates the similarity between concepts of semantic nets by using path lengths. Some studies [8] [9] have extended and used the conceptual distance for information retrieval. Resnik [10] has proposed an alternative similarity measure based on the concept of information content. Ganesan et al. [11] have presented new similarity measures in order to produce more intuitive similarity scores based on traditional measures.

On the other viewpoints, the reference [12] is surveyed. This survey [12] shows common architecture and general functionality as OBIE from various ontology-based information extraction researches. It consists of "information extraction module", "ontology generator", "ontology editor", "semantic lexicon" and some preprocessors. Their researchers are working for both of various researches on OBIE system implementation and research focusing on each module. In this paper, we will mainly introduce research on OBIE system implementation. The OBIE system implemented by Saggion et al. [13] operates the application of information extraction for a practical e-business. The system implemented by Wu et.al. [14] is one of the OBIE because of using the structure of "inforboxes" of Wikipedia as an ontology. The Cimiano et al. System [15] is a pattern-based approach to categorize instances with regard to ontology. OntoSyphon [16] uses the ontology to specify web searches that identify possible semantic instances, relations, and taxonomic information. The Maedche et al. system [17] is bootstrapping approach that allows for the fast creation of an ontology-based information extracting system. The TEXT-TO-ONTO Ontology Learning Environment [18] is based on a general architecture for discovering conceptual structures and engineering ontologies from text. SOBA [19] is a component for ontology-based information extraction from soccer Web pages. Embley [20] prefers the use of information-extraction ontologies as an approach that may lead to semantic understanding. Li et al. [21] proposed a hierarchical learning approach for information extraction. Hwang [22] approach is on dynamic ontologies that are automatically constructed from text data. OntoX [23] detect out-of-date constructs in the ontology to suggest changes to the user. Vulcain [24] identifies domain-specific terms and concepts, using syntactic information and an existing domain ontology. Vargas-Vera et al. [25] proposed Semantic Annotation Tool for extraction of knowledge structures from web pages through the use of simple user-defined knowledge extraction patterns. KIM [26] provides a mature and semantically enabled infrastructure for scalable and customizable information extraction. iDocument [27].

On the viewpoint of prediction, recently, some researchers work prediction methods. Eytan et al. [28] propose a prediction method of Internet user action from each personal history. Adam et al. [29] propose a prediction method of the user's health care from contents of posts twitter. Eric et al. [30] propose an association discovery between social media and economy.

Our method relatively compare with each conditional probability for effectual relationship discovery between each thing, event and phenomenon. In addition, we propose a new concept for discovery, an anteroposterior correlation.

3   Essence of Big Data Analytics

We have to focus on Big data analytics, which is completely different from such current data analytics as data mining technology. The key issues of Big data analytics are heterogeneous, continuity, and visualization.

3.1   Definition of Big Data

Recently, not only business people, but researchers are also focusing on Big data, which is defined by three Vs [31][32]:

- Volume: large amounts of data
- Variety: different forms of data, including traditional database, images, documents, and complex records
- Velocity: data content constantly changing through the absorption of complementary data collections and from streaming data from multiple sources

These 3V's definitions are on the viewpoint of infrastructures such as High Performance Computing and parallel distributed processing. These researches have finished. Because, big ICT companies such as Google, Amazon, Facebook, etc. operates these infrastructures as actual systems. What we have to consider is Big data infrastructures as a social problem. For example, Big data infrastructures need much electric power. The one of the important problems is how to operate Big data infrastructures without tragedy such as Fukushima. However, each big company is solving this problem. For example, Facebook will operate a Big data infrastructure by 100% wind-generated power as a system by 2016 [33]. It is impossible in Japan at least whose liberalization of electric power is not enough.

However, some people use keyword "Big data". We computer scientist should consider this to be needed. What are these needs? The keyword "Small data"[34] tells us a hint.

The definition of Small data is described by [34] as follows: Small data connect people with timely, meaningful insights (derived from big data and/or "local" sources), organized and packaged – often visually – to be accessible, understandable, and actionable for everyday tasks. This definition is similar to the advantages and sales talks of Big data. We consider that these needs mean, how to analyze non-schema data. In this context, the volume, including the 3V is not related, it may be large, or may be small, or whichever may be sufficient.

In order to analyze appropriately such above description, a system has to correctly map into cyber world from the real world. In the next section, I show the mapping from real world to cyber world.

3.2   Mapping from Real World to Cyber World

The one of the important elements for Big data analytics, including small data case such as [34] is mapped correctly in cyber world from the real world. Fig. 1 shows the relationship between real world and cyber world. Sensors aggregate real world situation as discrete data. However, the real world is continuous. In order to correctly analyze in cyber world, a system has to be analyzed by using continuous value. Therefore, fitting or interpreting is very important.

It is a very easy thing. For example, for listening to music, CDs and a CD player can be used. CDs have discrete sound data from the real world. Their music cannot be recognized without digital/analogue conversion by a CD player. This example shows us one important feature of Big data. Each piece only represents an instance of a certain state. Of course, higher sampling rates produce more correct data. However, the real world is a continuous place. Unless the data are continuous, many things cannot be discovered, like the example of a CD's music.

Other issues include which axis to interpolate. In the example of a CD's, music, we can interpolate the time axis by a digital/analogue converter. Depending on the information provided by the data, we do not know which interpolation uses place information or which uses the temperature of air/water information.

Finally, even though computer science researchers have researched approximation method from discrete to continuous values, they have never researched the selection or the creation of appropriate axes for continuous discovery. This is a critical theme of Big data analytics.

On the other aspects, we have very variety of sensors in the real world because of spreading mobile terminals and accurate low price sensor. Such sensors will continue increasing in number from now on also. Can we design a schema such as situation? It is impossible. Therefore, a system has to various data on non-schema. In addition, the system has to integrate these heterogeneous data.

Finally, most sensor data are fragmented. A web page is one of the data in the cyber world. We enjoyed these data as contents. Such data are becoming the past things. Current data are more fragments. For example, each tweet on twitter consists of short sentences or words; sensor data consists of a numerical value; etc. This means we cannot enjoy one data as contents. Therefore, we have to shift the paradigm of search engine. Currently, most search engines provide a list of data

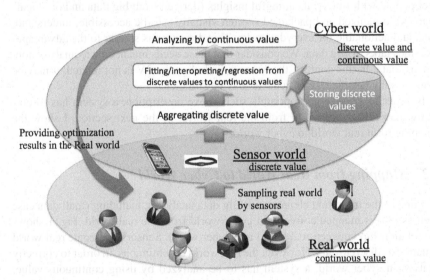

Fig. 1 Relationship between real world and cyber world. Sensors aggregate real world situation as discrete data. However real world is continuous. In order to correctly analyze in cyber world, a system has to be analyzed by using continuous value.

corresponding to the user's keywords by pattern matching. However, a user is beginning to lose user's interest by each data as contents. Almost users want to see the overview or trend of these data in the user's interest. It is not enough to see the overview or a trend of focused data by representation of a list.

We consider that it is important to create content by using fragment data in cyber world. It means that visualization is changing. By user's query, a system aggregate necessary data, analysis these data and create visualization along with analysis. Therefore, a system creates new contents for user corresponding to a user's query automatically. In the near future, a system will create actual something in the real world by 3D printers and other methods. This is a mutual mapping between real world and cyber world.

3.3 Three Primitive of Big Data Analytics

It is important to discover answers or clues for users in real time. A system has to create an appropriate schema from the data themselves given by user queries. Until now, data have been organized based on database schema. Currently, only various fragmentary data exist on the web.

This is a huge paradigm shift. We must create the schema and data structures that correspond to the processing required by users after they input queries. We have to shift the system from designing closed assumptions to opened assumptions.

Heterogeneity, continuity, and visualization are the most critical features of Big data analytics, which provides a scale and connection merits based on them. No current data analysis methods are based on opening assumptions. Big data analytics provide a new data analysis method based on opening assumptions. Below, we discuss the inconsistencies caused by continuing to use the current methods. Fig. 2 shows the relationship between the three elements (volume, variety, velocity) of Big data's definition and its analysis features (heterogeneity, continuity, visualization).

In Big data analytics, heterogeneity is different from the type to which the Big data definition belongs. The variety of Big data definitions include such content as images, sounds, documents, etc. Its heterogeneity includes such data fields as news, entertainment, technology, and science, all of which are semantic aspects.

Most big data come from sensors. For social media, each human action is part of the Big data from each human sensor. It is important to aggregate every second such massive and various sensor data in the Big data processing infrastructures. However, there is a restriction on the sampling rate. For more realistic analysis, we must approximate the aggregated discrete data to continuous value data, because the real world is continuous.

Visualization is important for Big data analytics for many different reasons. For example, a Google search provides a list that represents an appropriate data set. Is such a list of representations satisfactory? In Big data environments, not every piece of data can be clicked on and checked because we are not interested in such details. Since we want to identify the trends of the whole data set, visualization is a crucial issue for Big data analytics. We can use various visualization methods created by researchers.

We have to consider how to construct Big data analytics for heterogeneity, continuity, and visualization features. One critical issue is from what viewpoint to see Big data.

In this paper, we focus on the "heterogeneity" on the viewpoint of conditional probability. All things, events, and phenomenon are represented in occurrence probability. Our method discovers the relation of each interdisciplinary thing, event, and phenomenon without inconsistency by representing all in occurrence probability.

4 Anteroposterior Correlation Discovery Method

In this section, we represent our method, anteroposterior correlation discovery. The anteroposterior correlation means the correlation based on the time anteroposterior relation. The one of the features of our method is a measurement correlation by using conditional probability. We calculate the anteroposterior correlation relative by representing all in conditional probability.

In the section 4.1, we show our basic idea of the anteroposterior correlation. In the section 4.2, we propose a new evaluation index "degree of dependence" for he anteroposterior correlation relatively. In the section 4.3, we show an application example to query logs. A query log is the one of the important Big data. Especially,

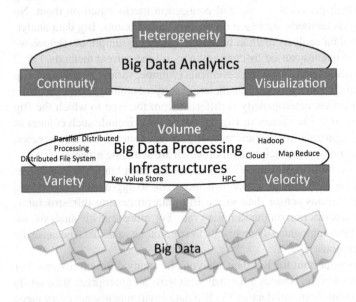

Fig. 2 Relationship among three Vs of Big data definition and Big data analytics definition: heterogeneity, continuity, and visualization

we show an example of application to Google Trends which provides the normalized values by each keyword and week.

4.1 Basic Idea of Our Method

Certain things, events or phenomenon represent in nuisance variable A, B, and C. Constraints, such as time and a place, etc., are expressed as parameters time, place, etc. The relationship event A and phenomenon B in the case of constraints in time and place represents as follows a formula:

$$p(A|B, time, place) = \frac{n_{A \cap B, time, place}}{n_{B, time, place}}, \tag{1}$$

where $n_{A \cap B, time, place}$ is the number of co-occurence of event A and phenomenon B ($A \cap B$) in constraints of a *time* and a *place* and $n_{B, time, place}$ is the number of phenomenon in constraints of a *time* and a *place*. This formula means the probability of occurrence, event A given phenomenon B in the case of constraints a time and a place. In this case, as for the time relation, the phenomenon B has occurred more previously than event A. In addition, the phenomenon B correlates to event A. We define such relation between A and B as anteroposterior correlation.

In this formula, we describe the parameter explicitly. We can add more parameters as constraints. In this paper, generally, we do not describe the parameter such as follows the formula:

$$p(A|B, time, place) = \frac{n_{A \cap B, time, place}}{n_{B, time, place}} \cong p(A|B). \tag{2}$$

We observe the value of this anteroposterior correlation relatively. For example, you would like to investigate whether an anteroposterior correlation, of which is higher in Phenomenon B between event A and event C. You only compare value $p(A|B)$ with value $p(C|B)$. The higher value presupposes that anteroposterior correlation is higher. By processing these comparisons repeatedly, we can find a good anteroposterior correlation in the given constraints.

This is the one of the points for our method. All things, event and phenomena represent in conditional probability and these values compare relatively like similarity value. Usually, although a probability value is an absolute value, our method can find out a more capable relation by comparing with other probability.

4.2 Degree of Dependence

You may say what difference it is from co-occurrence. It is a very important point. We focus on dependency.

For example, there are event A and event B. The event A expresses the phenomenon which causes a traffic accident. The event B expresses the phenomenon which drinks alcohol. In Japan, $p(A, B)$ is a low value. Since the penal regulations of drunken driving are severe, those who drink alcohol and drive are rare. However, $p(A|B)$ is a high value because When alcohol is drunk, he/she will get drunker and driving a car will also become very dangerous.

This example shows the dependence about B of the anteroposterior correlation between A and B. We call it a degree of dependence.

The following formula holds when event A and event B are independent:

$$p(A,B) = p(A)p(B). \tag{3}$$

The following formula holds when event A and event B are not independent:

$$p(A,B) = p(A|B)p(B). \tag{4}$$

By two formulas, we focus on the ratio of $p(A)$ and $p(A|B)$. The ratio is a good evaluation index for dependency. Therefore, the degree of dependence about B of the anteroposterior correlation between A and B represents as follows a formula:

$$degreeOfDependence_{p(B)p(A|B)} = \frac{p(A|B)}{p(A)}. \tag{5}$$

The higher evaluation index value presupposes that the degree of dependence about B of the anteroposterior correlation between A and B is higher. This shows that a possibility that event B has caused event A is higher. That is, the higher evaluation index value presupposes better anteroposterior correlation.

Both Comparison of the conditional probability described in section 4.1 and the evaluation index of the degree of dependence are used, and the system based on our method extract appropriate anteroposterior correlation.

4.3 Application to Query Log

We apply our method to a query log. The nuisance variable A expresses the phenomenon which inputs the query of a certain phenomenon X. The nuisance variable B shows that a certain phenomenon Y occurs. The t_B is the time when the phenomenon B occurs.

In this situation, the probability $p(A)$ represents as follows a formula:

$$p(A) = \frac{\int_{time} n_A(time)}{\int_{time} N_A(time)}, \tag{6}$$

where $N(time)$ is all the number of times of all the query of all the time, $n_A(time)$ is the number of times of the query phenomenon X. The probability $p(A|B)$ represents as following formula:

$$p(A|B) = \frac{p(A,B)}{p(B)} = \frac{\int_{t_B}^{t_{const}} n_A(time)}{p(B)\int_{t_B}^{t_{const}} N_A(time)}, \tag{7}$$

where t_B is the time of occurrence of phenomenon Y and t_{const} is a time window after t_B given as the one of the parameters.

Here, these values derive for comparisons. In this case, the nuisance variable B, $\int N(time)$, t_B and t_{const} are fixed. Therefore, we can simplify above two formulas:

$$p(A) = \frac{\int_{time} n_A(time)}{\int_{time} N_A(time)} \propto \int_{time} n_A(time), \tag{8}$$

$$p(A|B) = \frac{p(A,B)}{p(B)} \propto p(A,B) \propto \int_{t_B}^{t_{const}} n_A(time). \tag{9}$$

Fig. 3 shows the value of $p(A)$ and $p(A|B) \propto p(A,B)$ on the time axis and the number of times for query of a certain phenomenon X. A slash part is a value of $p(A)$. A colored part is a value of $p(A|B) \propto p(A,B)$.

These simplifications have a fatal fault. The value of anteroposterior correlation and co-occurrence are same. We cannot extract the actual anteroposterior correlation. Therefore, we compare and extract anteroposterior correlations by calculating the degree of dependence as shown in the section 4.2.

By calculating the degree of dependence, we extract appropriate anteroposterior correlations.

We show a simple example, as shown in Fig. 4. We relatively compare the case of A_1, A_2, A_3, and A_4.

We cannot calculate the degree of dependence each item A_1, A_2, A_3, and A_4, because their values are not normalized. Without normalization, the item which is not meant will become a big value. For example, A_3 is a bigger value without nor-

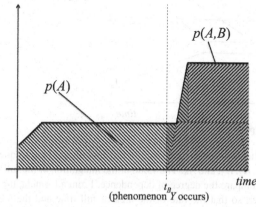

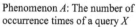

Fig. 3 The graph for time axis and the number of occurrence times of a query X. A slash part is a value of $p(A)$. A colored part is a value of $p(A,B)$.

malization than A_2. In this paper, the value of the integration of the full time of A_i normalizes to 1. A more effective normalization system will be in the future work.

Next, we compare the degree of dependence. From a formula, the degree of dependence becomes higher, so that the rate of the area of full time and the area after t_B is higher. Therefore, we can see that A_2 and B are more appropriate anteroposterior correlated than A_1 and B.

However, in the case of A_4, although there is a dependency of phenomenon Y, the degree of dependence is a low value. Therefore, we have to calculate the degree of dependence in the case of at the same time.

5 Preliminary Experiment

5.1 Experiment System Setting

We focus on the Great East Japan Earthquake. Our purpose is an extraction of appropriate anteroposterior correlation of the Great East Japan Earthquake.

We use Google Trends as query log. The Google Trends provides the number of times for a keyword given by a user in each week or each month. Although this data is secondary data, it can be seen as Big data. A query log is an important Big data. In this paper, we only use Google Trends as a query data. Therefore, our method can use the actual query log such as other search engine or retrieval system in intranet.

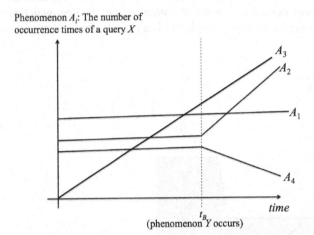

Phenomenon A_i: The number of occurrence times of a query X

A_3
A_2
A_1
A_4
time
t_B
(phenomenon Y occurs)

Fig. 4 A simple example of the degree of dependence in the case of applying to query log. First, we have to normalize A_1, A_2, A_3, and A_4. The value of the integration of the full time of A_i normalizes to 1. Next, we compare the degree of dependence. From a formula, the degree of dependence becomes higher, so that the rate of the area of full time and the area after t_B is higher. However, in the case of A_4, although there is the dependency of phenomenon Y, the degree of dependence is a low value. Therefore, we have to calculate the degree of dependence in the case of "not A_i" at the same time.

5.2 Experimental Result

Fig. 5 shows an Experimental result. We calculate the degree of dependence of the Great East Japan Earthquake (2011.3.11) for anteroposterior correlations between each disease and the Great East Japan Earthquake. For comparison, we calculate the degree of dependence of the Great East Japan Earthquake for anteroposterior correlations between SNS and the Great East Japan Earthquake. We regard this "SNS" value as a baseline level, because almost Japanese used SNS in this earthquake.

We can see the strong anteroposterior correlations between this earthquake and "thyroid cancer," "autonomic dysregulation," "disorder of emotion," "trauma," "coronary arteriopathy," "common cold," "bronchitis." This result shows having the influence of radioactivity rather than the influence of earthquake directly.

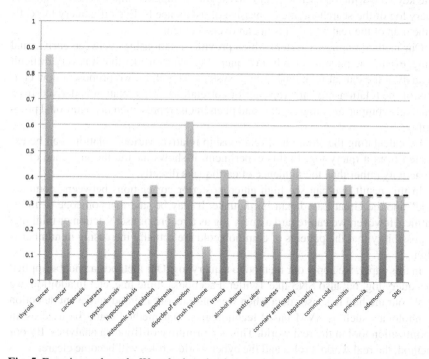

Fig. 5 Experimental result. We calculate the degree of dependence of the Great East Japan Earthquake (2011.3.11) for anteroposterior correlations between each disease and the Great East Japan Earthquake. For comparision, we calculate the degree of dependence of the Great East Japan Earthquake for anteroposterior correlations between SNS and the Great East Japan Earthquake. We regard this "SNS" value as a baseline level, because almost Japanese used SNS in this earthquake. We can see the strong anteroposterior correlations between this earthquake and "thyroid cancer," "autonomic dysregulation," "disorder of emotion," "trauma," "coronary arteriopathy," "common cold," "bronchitis." This result shows having influence of radioactivity rather than the influence of earthquake direct.

In this result, we can see high anteroposterior correlations between "common cold" and the Great East Japan Earthquake. It may just happen that way. However, we can consider that common cold increased as a result of physical strength's weakening from refuge from an earthquake, or the influence of radioactivity. In this investigation [35], after 93% of Japanese people answer they search their condition on the Internet mostly, after that they go to a hospital. It will be clear that common cold becomes epidemic after the Great Eastern Japan Earthquake.

By this experimental result, our method discovers anteroposterior correlations easily. This discovery gives us new perceptions.

5.3   Discussion

The keyword which a user wants to investigate is inputted into a search engine. A query log of the search engine is massive of user's needs. Even if a query log calls it the map of the real world, it is not an overstatement.

Our method realizes a system which provides relationships between each actual thing, event, and phenomenon from a query log. We consider that it is very difficult to analyze the causal relationship by a system. Therefore, we propose a new concept of the relationship "anteroposterior correlation." This relation is derived from relatively comparing things, events and phenomena represented in conditional probability.

By calculating the strength of relationship relative, various relations can be extracted from a query log. In this experiment, it shows having the influence of radioactivity rather than the influence of earthquake directly.

In this result, we can see high anteroposterior correlations between "common cold" and the Great East Japan Earthquake. It may be said that someone is unscientific. However, we can regard a query log as a record of user's action or thinking. A possibility that the patients of common cold are actually increasing in number is high.

In this paper, we apply our method to a query log. Our method can not be applied to a query log, but also SNS like Twitter, Facebook's time line, etc. In addition, we apply our method to sensor data when we combine our method and event detection technologies such as SVM based technologies. Our method could be used as an optimization tool in the real world. This is a primitive of Big data analytics. By our method, the real world's roles and the cyber world's roles will become clearer.

6   Conclusion

We proposed a new concept, anteroposterior correlation. The anteroposterior correlation means the correlation based on the time anteroposterior relation. In this paper, we represented a new discover method of anteroposterior correlation between each heterogeneous thing, events and phenomena. We discover a correlation in consideration of the continuity of time. By our method, we effectively discover relationship between heterogeneous things, events and phenomena. The one of the features of our method is a measurement correlation by using conditional probability. We calculate the anteroposterior correlation relative by representing all in conditional probability.

We introduced our ethic of essences for Big data analytics. Heterogeneity, continuity, and visualization are the most critical features of Big data analytics, which provides a scale and connection merits based on them. We have to shift the system from designing closed assumptions to opened assumptions.

In addition, we also verified the effectiveness of our method by preliminary experiment. We used Google Trends. The Google Trends can be regarded as query logs. This is the one of the Big data. Therefore, our method provides the anteroposterior correlation from Big data. This experiment shows that we realize the one of the methods for decision mining of our method.

By applying our method, we will realize the system which can do risk aversion, the cause unfolding, and the phenomena prediction in the near future. The person harnesses the past experience and does prediction and prevention. Our system can do the same things like this. That is, we can predict and prevention of Big data. Therefore, we realize the one of the methods for decision mining.

References

1. Salton, G., Wong, A., Yang, C.S.: A vector space model for automatic indexing. Magazine Communications of the ACM CACM Homepage Archive 18(11), 613–620 (1975)
2. Deerwester, S., Dumais, S.T., Furnas, G.W., Landauer, T.K., Harshman, R.: Indexing by latent semantic analysis. Journal of the American Society for Information Science 41(6), 391–407 (1990)
3. Kitagawa, T., Kiyoki, Y.: A mathematical model of meaning and its application to multidatabase systems. In: RIDE-IMS 1993: Proceedings of the 3rd International Workshop on Research Issues in Data Engineering: Interoperability in Multidatabase Systems, pp. 130–135 (1993)
4. Kiyoki, Y., Kitagawa, T., Hayama, T.: A metadatabase system for semantic image search by a mathematical model of meaning. SIGMOD Rec. 23(4), 34–41 (1994)
5. Takano, K., Kiyoki, Y.: A superordinate and subordinate relationship computation method and its application to aerospace engineering information. In: ACST 2007: Proceedings of the Third Conference on IASTED International Conference, Anaheim, CA, USA, pp. 510–516 (2007)
6. Miller, G.A., Beckwith, R., Fellbaum, C., Gross, D., Miller, K.J.: Introduction to WordNet: An on-line lexical database. Journal of Lexicography 3(4), 235–244 (1990)
7. Rada, R., Mili, H., Bicknell, E., Blettner, M.: Development and application of a metric on semantic nets. IEEE Transactions on Systems, Man and Cybernetics 19(1), 17–30 (1989)
8. Kim, Y., Kim, J.: A model of knowledge based information retrieval with hierarchical concept graph. Journal of Documentation 46(2), 113–136 (1990)
9. Lee, J., Kim, M., Lee, Y.: Information retrieval based on conceptual distance in is-a hierarchies. Journal of Documentation 49(2), 188–207 (1993)
10. Resnik, P.: Using information content to evaluate semantic similarity in a taxonomy. In: IJCAI: Proceedings of the International Joint Conference on Artificial Intelligence, pp. 448–453 (1995)
11. Ganesan, P., Garcia-Molina, H., Widom, J.: Exploiting hierarchical domain structure to compute similarity. ACM Trans. Inf. Syst. 21(1), 64–93 (2003)

12. Wimalasuriya, D., Dou, D.: Ontology-based information extraction: An introduction and a survey of current approaches. Journal of Information Science 36(3), 306–323 (2010)
13. Saggion, H., Funk, A., Maynard, D., Bontcheva, K.: Ontology-Based Information Extraction for Business Intelligence. In: Aberer, K., et al. (eds.) ASWC/ISWC 2007. LNCS, vol. 4825, pp. 843–856. Springer, Heidelberg (2007)
14. Wu, F., Hoffmann, R., Weld, D.S.: Information extraction from Wikipedia: moving down the long tail. In: Proceedings of the 14th ACM SIGKDD International Conference on Knowledge Discovery and Data Mining (KDD 2008), pp. 731–739. ACM, New York (2008), http://doi.acm.org/10.1145/1401890.1401978, doi:10.1145/1401890.1401978
15. Cimiano, P., Handschuh, S., Staab, S.: Towards the self-annotating web. In: Proceedings of the 13th International Conference on World Wide Web (WWW 2004), pp. 462–471. ACM, New York (2004), http://doi.acm.org/10.1145/988672.988735, doi:10.1145/988672.988735
16. McDowell, L.K., Cafarella, M.: Ontology-Driven Information Extraction with OntoSyphon. In: Cruz, I., Decker, S., Allemang, D., Preist, C., Schwabe, D., Mika, P., Uschold, M., Aroyo, L.M. (eds.) ISWC 2006. LNCS, vol. 4273, pp. 428–444. Springer, Heidelberg (2006)
17. Maedche, A., Neumann, G., Staab, S.: Bootstrapping an ontology-based information extraction system. In: Szczepaniak, P.S., Segovia, J., Kacprzyk, J., Zadeh, L.A. (eds.) Intelligent Exploration of the Web, pp. 345–359. Physica-Verlag GmbH, Heidelberg (2003)
18. Maedche, A., Staab, S.: The Text-to-Onto Ontology Learning Environment. In: Software Demonstration at the 8th International Comference Conceputual Structures. Springer, Berlin (2000)
19. Buitelaar, P., Siegel, M.: The Text-to-Onto Ontology Learning Environment. In: Proceedings of the 5th International Conference on Language Resources and Evaluation, pp. 2321–2324 (2006)
20. Embley, D., David Embley, W.: Toward semantic understanding: an approach based on information extraction ontologies. In: Schewe, K.-D., Williams, H. (eds.) Proceedings of the 15th Australasian Database Conference (ADC 2004), vol. 27, pp. 3–12. Australian Computer Society, Inc., Darlinghurst (2004)
21. Li, Y., Bontcheva, K.: Hierarchical, perceptron-like learning for ontology-based information extraction. In: Proceedings of the 16th International Conference on World Wide Web (WWW 2007), pp. 777–786. ACM, New York (2007), http://doi.acm.org/10.1145/1242572.1242677, doi:10.1145/1242572.1242677
22. Hwang, C.: Incompletely and imprecisely speaking: Using dynamic ontologies for representing and retrieving information. In: Proceedings of the 6th International Workshop on Ontology-Based Information Extraction System, Kaiserslautern, Germany (1999)
23. Yildiz, B., Miksch, S.: ontoX - A Method for Ontology-Driven Information Extraction. In: Gervasi, O., Gavrilova, M.L. (eds.) ICCSA 2007, Part III. LNCS, vol. 4707, pp. 660–673. Springer, Heidelberg (2007)
24. Todirascu, A., Romary, L., Bekhouche, D.: Vulcain - An Ontology-Based Information Extraction System. In: Andersson, B., Bergholtz, M., Johannesson, P. (eds.) NLDB 2002. LNCS, vol. 2553, pp. 64–75. Springer, Heidelberg (2002)
25. Vargas-Vera, M., Motta, E., Domingu, J., Shum, S., Lanzoni, M.: Knowledge extraction by using an ontology-based annotation tool. In: Proceedings of the workshop on knowledge markup and semantic annotation, USA, ACM Press, New York (2001)

26. Popov, B., Kiryakov, A., Ognyanoff, D., Monov, D., Kirilov, A.: KIM–a semantic platform for information extraction and retrieval.? Natural Language Engineering 10(3-4), 375–392 (2004)
27. Adrian, B., Hees, J., van Elst, L., Dengel, A.: iDocument: Using Ontologies for Extracting and Annotating Information from Unstructured Text. In: Mertsching, B., Hund, M., Aziz, Z. (eds.) KI 2009. LNCS, vol. 5803, pp. 249–256. Springer, Heidelberg (2009)
28. Adar, E., Weld, D.S., Bershad, B.N., Gribble, S.D.: Why We Search: Visualizing and Predicting User Behavior. WWW (2007)
29. Sadilek, A., Kautz, H., Silenzio, V.: Predicting Disease Transmission from Geo-Tagged Micro-Blog Data. AAAI (2012)
30. Gilbert, E., Karahalios, K.: Widespread Worry and the Stock Market. AAAI (2010)
31. Berman, J.: Principles of Big Data Preparing. Elsevier / Morgan Kaufmann (2013)
32. Minelli, M., Chambers, M., Dhiraj, A.: Big Data, Big Analytics: Emerging Business Intelligence and Analytic Trends for Today's Businesses. Wiley (2013)
33. Finley, K.: Facebook Says Its New Data Center Will Run Entirely on Wind, WIRED (2013),
    http://www.wired.com/wiredenterprise/2013/11/facebook-iowa-wind/
34. Greenberg, P.: 10 Reasons 2014 will be the Year of Small Data, ZDNet. (2013),
    http://www.zdnet.com/10-reasons-2014-will-be-the-year-of-small-data-7000023667/
35. http://japan.internet.com/wmnews/20100628/3.html

Increasing the Accuracy of Software Fault Prediction Using Majority Ranking Fuzzy Clustering

Golnoush Abaei and Ali Selamat

Abstract. Although many machine-learning and statistical techniques have been proposed widely for defining fault prone modules during software fault prediction, but this area have yet to be explored as still there is a room for stable and consistent model with high accuracy. In this paper, a new method is proposed to increase the accuracy of fault prediction based on fuzzy clustering and majority ranking. In the proposed method, the effect of irrelevant and inconsistent modules on fault prediction is decreased by designing a new framework, in which the entire project's modules are clustered. The obtained results showed that fuzzy clustering could decrease the negative effect of irrelevant modules on accuracy of estimations. We used eight data sets from NASA and Turkish white-goods software to evaluate our results. Performance evaluation in terms of false positive rate, false negative rate, and overall error showed the superiority of our model compared to other predicting strategies. Our proposed majority ranking fuzzy clustering approach showed between 3% to 18% and 1% to 4% improvement in false negative rate and overall error respectively compared to other available proposed models (ACF and ACN) in at least half of the testing cases. The results show that our systems can be used to guide testing effort by prioritizing the module's faults in order to improve the quality of software development and software testing in a limited time and budget.

Keywords: Software fault prediction, Fuzzy clustering, False negative rate (FNR), False positive rate (FPR).

Golnoush Abae
Software Engineering Research Group, Faculty of Computing,
University Technology Malaysia (UTM), Johor Baharu, 81310, Johor, Malaysia
e-mail: golnoosh.abaee@gmail.com

Ali Selamat
UTM-IRDA Digital Media Centre, K-Economy Research Alliance UTM &
Software Engineering Research Group, Faculty of Computing,
University Technology Malaysia (UTM), Johor Baharu, 81310, Johor, Malaysia
e-mail: aselamat@utm.my

© Springer International Publishing Switzerland 2015   179
R. Lee (ed.), *SNPD*,
Studies in Computational Intelligence 569, DOI: 10.1007/978-3-319-10389-1_13

1 Introduction

Since software projects play major role in nowadays industry, the accurate estimating of the software development cost is very important. According to the Standish group report in 2009 [1], just 32% of software projects were on time and on cost in 2009, 44% of the projects were in challenged mode and 24% of projects had been cancelled. Designing, developing, testing, and all aspects of the software projects are affected by the relevant estimations and predictions. Software testing is known as a major factor in increasing the development cost. Faulty modules cause significant risk by decreasing customer fulfillment and by increasing the testing, development, and maintenance costs. Early detection of fault-prone software components could enable verification experts and testers to concentrate their time and resources on the problematic areas of the system under development. During the recent decades, many methods for the software fault prediction have been proposed. Selecting a method as the best one seems to be impossible because the performance of each method depends on the various factors such as different software measurement metrics, available information, machine-learning techniques and so on. However, the main aim of all methods is presenting the accurate results.

Area of software fault prediction still poses many challenges and unfortunately, none of the techniques proposed within the last decade has achieved widespread applicability in the software industry. This is due to several reasons including the lack of software tools to automate this prediction process, the unwillingness to collect the fault data, and other practical problems [2, 3].

Soft computing is a term that is recently become popular in the general area of prediction. It is a field within computer science that is characterized by the use of inexact solutions. Soft computing differs from conventional (hard) computing in that, unlike hard computing, soft computing deals with imprecision, uncertainty, partial truth, and approximation to achieve practicability, robustness, and low solution cost [4]. As such, it forms the basis of a considerable amount of machine learning techniques. Components of soft computing include neural networks, support vector machines, fuzzy logics, evolutionary computation and so on. Although many papers have been published on software defect prediction techniques, machine-learning and soft computing approaches have yet to be fully explored.

In fault prediction process, previous reported faulty data along with distinct metrics identify the fault-prone modules. However, outliers and irrelevant data in training set can lead to the imprecise estimations. In fact, in many engineering problems, we encounter vagueness in information and uncertainty in training sets, so as these phenomena cause, we could not reach to certain results for our proposed solution. Our system, models the input information's vagueness through fuzzy clusters and fault prediction is done based on majority ranking of three most similar fuzzy clusters with the test data. This system provides more accurate results compared to existing methods based on different classification techniques.

Based on our proposed model we construct 3 research questions that are shown as follows:

- **RQ1:** Is fuzzy clustering with majority ranking performing better than two well-performed learning methods in fault prediction modeling namely naïve bayes and random forest?

- **RQ2:** Is fuzzy clustering with majority ranking performing better than two well-performed learning methods in fault prediction modeling namely naïve bayes and random forest when two-stage outlier removal is applied on data sets?

- **RQ3:** How our proposed model performed when two different sets of datasets are used for training and testing process?

The remainder of this paper continues with section 2, where a brief discussion on related works is presented. Fuzzy clustering is reviewed in section 3. Section 4 contains our proposed method. Experimental descriptions are presented in section 5. Experimental results and analysis are described in section 6, and finally, we summarize this paper in section 7.

2 Related Works

According to Catal[5], software fault prediction became one of the noteworthy research topics since 1990 and it includes two recent and comprehensive systematic literature reviews [2,6]. The prediction techniques use approaches that originated from the field of either statistics or machine learning. Some of these techniques are genetic programming [7], decision trees [8] neural network [9], naïve bayes[10], case-based reasoning [11], fuzzy logic [12] and the artificial immune recognition system algorithms in [13,14,15]. As the number of related works in this area is too much, we just presented some of them in this section.

Menzies et al. [10] conducted several experiments based on different data mining algorithms with method level metrics on public NASA datasets. They evaluated their work with probability of false alarm (PF) and probability of detection (PD), and balance. They reported the best performer as naïve bayes and they used log-transformation with Info-Gain filters before applying the algorithms. They claimed that the best algorithm changes according to the dataset characteristics and numerous experiments should be performed for a robust prediction model. They also argued that since some models with low precision performed well, using it as a reliable parameter for performance evaluation is not recommended. Although Zhang et al. [16] criticized the paper but Menzies et al. defended their claim in [17].

Mahaweerawat et al. [18] presented a new approach for predicting software faults by means of fuzzy clustering and radial basis function techniques. They applied the radial-basis function network after they used fuzzy subtractive clustering to divide historical data into clusters in order to predict faults that occurred in the component residing in each cluster. In 2007 [19], they performed a similar experiment by using self-organizing map as a classifier instead of fuzzy approach. In both experiments, the claimed that they could predict software faults reasonably accurate but unfortunately very few experimental details were provided by them in both papers.

Zhong et al. [20,21] applied clustering along with expert-based approach to solve fault prediction problem. They used k-means and neural-gas techniques for clustering different real data sets and then an expert decided whether each cluster representative should be labeled as faulty or non-faulty. After analyzing the results in terms of the overall error rate, they affirmed that k-means clustering-based approached performed slightly better than neural-gas-based approach on large data sets.

Yuan et al. [22] used fuzzy subtractive clustering with module-order modeling in order to build prediction model. First fuzzy subtractive clustering was used to predict the number of faults, and then module-order modeling was applied to predict whether modules were faulty or not. Based on the case study, it was found that proposed approach could classify modules that will likely have faults.

Catal and Diri [13] focused on the high-performance fault predictors based on machine learning such as random forests and algorithms based on artificial immune systems on public NASA datasets. They reported that random forests provides the best prediction performance for large datasets and naive bayes is the best prediction algorithm for small datasets in terms of the area under receiver operating characteristics curve (AUC).

Alan and Catal [23] proposed an outlier detection approach using metrics thresholds and class labels to identify class outliers. They evaluated their approach on public NASA datasets. They stated that there proposed outlier detection method improves the performance of robust fault prediction models based on naïve bayes and random forests algorithms.

Rodriguez et al. [24] investigated two well-known subgroup discovery algorithms, the SD algorithm, and the CN2-SD algorithm to obtain rules that identify defect prone modules. The experiments performed on object-oriented metrics datasets from Eclipse repository showed that the EDER-SD algorithm performs well in most cases when compared to three other well-known SD algorithms.

3 Fuzzy Clustering

Clustering algorithms group the modules according to similarity of their software attributes. In fact, program modules with similar attributes are clustered together as they have similar quality characteristics; furthermore, dissimilarity of data located in separate clusters should be as high as possible. Proper data clustering technique will enhance not only the efficiency of the training process, but also the performance of the model predictability precision. Accurate predictions obtained from such a good reliability model will be favorable toward higher software process

efficiency and product quality [19]. Several parameters such as connectivity, inten-
sively and distance among data characteristics determine the level of similarity.
Usually in clustering methods data element belongs to exactly one cluster, which is
famous as hard clustering, however, among them, a soft clustering method that is
called fuzzy clustering calculates the relativity of each module $(X = x_1, x_2, ..., x_n)$ to
the specified clusters $(C = c_1, c_2, ..., c_c)$ with membership values $(M = m_1, m_2, ...,$
$m_n)$ varies from zero to one. In this method, data elements belong to one or more
clusters at the same time. The C-means clustering is one of the most important
fuzzy clustering techniques developed in 1973 [25] and improved in 1981 [26].
Variety of different application has used this method to solve their problems. In this
method, the final aim is to minimize a target function as shown in Eq. 1.

$$J_m = \sum_{i=1}^{N} \sum_{j=1}^{C} u_{ij}^m \left\| x_i - c_j \right\|^2 , \; 1 \leq m < \infty \tag{1}$$

u_{ij} is the membership degree of x_i from the center of cluster j (c_j), and $\|x_i - c_j\|$ is the
difference expressing the similarity between data (x_i) and the center of cluster j (c_j).

3.1   C-means Clustering Algorithm

In C-means clustering, first a set of random initial membership values $(U^{(0)} = u_{ij})$ are
generated from each data module x_i for each cluster c_j. Then center vector of each
cluster is calculated based on Eq. 2 for k number of times. After that $u^{(k)}$ and $u^{(k+1)}$ is
updated according to Eq. 3. And finally if difference between $u^{(k)}$ and $u^{(k+1)}$ is less
that the threshold, the iteration stops, otherwise, new cluster's centers are employed
based on Eq. 2

$$c_j = \frac{\sum_{i=1}^{N} u_{ij}^m \cdot x_i}{\sum_{i=1}^{N} u_{ij}^m} \; , \; c^{(k)} = \left[c_j\right] with U^{(k)} \tag{2}$$

$$u_{ij} = \frac{1}{\sum_{k=1}^{C} \left(\frac{\|x_i - c_j\|}{\|x_i - c_k\|} \right)^{\frac{2}{m-1}}} \tag{3}$$

4   Proposed Method

Clustering of the software projects is the key part of estimation method proposed in
this section. To overcome the diversity and inconsistency of the projects collected
in a dataset, it is required to separate the outliers and irrelevant projects from other
ones. The modules clustering can increase the consistency of modules by putting
similar modules in the same clusters. Instead of having a dataset, which includes
numerous irrelevant and inconsistent modules, there will be several subsets
comprising of consistent and similar modules. Clustering process is performed by
analyzing the modules features to discriminate the most similar modules and

putting them in the separate clusters. This is the first step in the proposed method. C-means clustering technique (explained in Section 3.1) follows the principles of fuzzy logic, which makes it an appropriate method to deal with the vagueness and complexity of software modules features. It is able to analyze both numerical and categorical features, and select the most suitable group for each modules based on the membership degree. Moreover, it is a robust technique against outlier presence. Therefore, it is selected to be used for software measurement modules clustering in the first stage of the proposed method.

The proposed method is organized in two main phases, which are training and testing stages. In the training stage, the structure of the proposed method is configured and in the testing stage, the software faultiness is predicted. Training and testing stages are described as follows.

4.1 Training Stage

As in is shown in Fig.1, first 66% of any project is selected for the training purpose. Then fuzzy clustering is applied on the training dataset and the values of the cluster's centers are recorded. In addition, all produced clusters are also saved along with their center's weights. We should mention that this experiment is repeated three times to make sure the different combination of testing and training datasets are examined. We also trained and tested our proposed model based on selecting datasets from different available projects in either NASA or Turkish white-goods manufacturer developing embedded controller datasets [27].

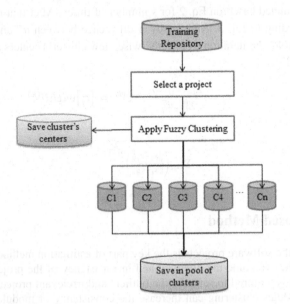

Fig. 1 Training stage of the proposed model

Since the threshold values for only seven different metrics were established by Integrated Software Metrics, Inc. (ISM) [28] as a deciding factor about faultiness of the modules, we clustered the NASA set as well as Turkish set (refer to 5.1) based on these thresholds. According to our experimental results, all selected datasets performed best with choosing the number of clustered as seven and six for NASA and Turkish sets respectively. The reason why Turkish set is clustered based on six clusters is that, one of the measurement's metrics did not identified for Turkish dataset. The metrics that clustering has been done based on them are line of code, cyclomatic complexity, unique operator, unique operand, total operator, total operand, and essential complexity. Essential complexity is not available for Turkish dataset.

4.2   Two Stage Outlier Removal

In this part, our motivation was to use outlier detection methods based on Alan and Catal [23] and that depend on the usage of software metrics thresholds and fault data (depicted as faulty or not faulty in the software measurement dataset).Outlier removal is done in two stages. At first, data object are marked as outlier and eliminated if five or six metrics exceed their corresponding metrics thresholds and if the data object's class label is not-faulty. After this stage, data objects are removed from data set if all the measurement metrics are below metrics thresholds and if the class label is faulty.

4.3   Testing Stage

According to Fig. 2, testing stage consists of three main steps. At first, a module from test dataset is selected. Then similarity between that selected module and cluster's centers are calculated based on Euclidean distance to specify the three most similar clusters to the selected test module. Second, after these three clusters are identified, again, the similarity between each of these clusters and tested module is computed and label of the test module (faulty/non-faulty) is determined according to the label of the most similar modules in each of the three selected clusters with the test module. The final decision about the label of the test module is made based on the majority ranking of the labels of the selected modules, which has the most similarity with test module. For example if two out of three of these similar selected modules from the clusters predict the module's label as faulty, the test module is labeled as faulty.

5   Experimental Description

This section describes all information required for conducting the experiments such as dataset selection and performance evaluation criteria.

5.1   Dataset Description

Eight data sets [27] are selected which are different in number of rows and defects rate.   These datasets are divided into two sets, NASA and Turkish set. NASA sets including CM1, KC1, KC2, JM1, and PC1 are used for comparison study with Alan and Catal [23] experiments and Turkish set with AR3, AR4, and AR5 are selected to show the generality of the proposed model.

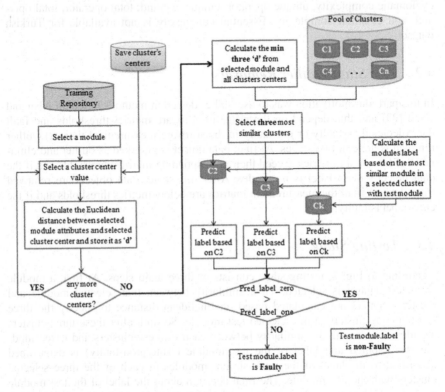

Fig. 2 Testing stage of the proposed model

The CM1 data set contains 498 modules and 10% of them are defected; this data belongs to NASA spacecraft instrument project. PC1 belongs to flight software for earth orbiting satellite; it has 1109 numbers of rows and 7% of the data are defected. KC1 with 2107 models which 15% of them are defected belongs to storage management project for receiving and processing ground data. The KC2 project belongs to the science data processing unit of storage management system used for receiving and delivering ground data and it contains 522 modules, of which 20% of them are defected. Finally, the largest dataset JM1 with 10,885 rows belongs to real time predictive ground system project; 19% of these data are defected.

AR3, AR4, and AR5 belong to Turkish white-goods manufacturer developing embedded controller software. AR3 has 63 modules of which 13% of it is faulty; AR4 is a larger data set with 107 rows and 19% faulty modules. The AR5 data set has only 36 modules and 22% of them are faulty.

5.2 Performance Evaluation Metrics

After specifying each module's label in testing phase, each evaluation metrics is calculated based on the confusion matrix. If a module's label predicted as non-faulty but the actual label is faulty, we get the condition of false negative (FN). On the other hand, if the non-faulty modules was labeled as faulty, we call it false positive (FP). If the faulty module predicted as faulty and non-faulty module predicted as non-faulty, they are called true positive (TP) and true negative (TN) respectively. FNR (false negative rate) is the percentage of modules that were actually faulty but there were predicted as non-faulty. In contrast, FPR (false positive rate) is the percentage of modules that were actually non-faulty but there were predicted as faulty. FNR, FPR, and errors need to be as small as possible in all the experiments. The following equations are used to calculate FNR, FPR, and overall error rate.

$$OverallErrorRate = \frac{FN+FP}{TP+FN+FP+TN} \tag{4}$$

$$FPR = \frac{FP}{FP+TN} \tag{5}$$

$$FNR = \frac{FN}{TP+FN} \tag{6}$$

5.3 Hypothesis Formulation

Each hypothesis listed as follows is identified according to our research question:

- **HP1:** Our proposed method performs better than naïve bayes and random forest in identifying fault prone and not faults prone modules. (Null Hypothesis: Our approach performs as well as naïve bayes and random forest in identifying fault prone and not faults prone modules.)
- **HP2:** Our proposed approach performed better than naïve bayes and random forest when two-stage outlier removal is applied. (Null Hypothesis: Our approach cannot outperform naïve bayes and random forest when two-stage outlier removal is applied.)
- **HP3:** Our proposed model performed well when it is applied on two different sets of training and testing datasets. (Null Hypothesis: Our proposed model does not perform well when it is applied on two different sets of training and testing datasets.)

6   Experimental Results and Analysis

In order to evaluate our work, we used two well-performed learning methods based on the literature reviews namely random forest and naïve bayes [2, 13, 29]. This section composed of two sets of experiments. Tables 1 to 7 present the prediction error analysis for the proposed fuzzy approach as compared to other approaches, namely, naïvebayes (NB), random forest (RF), Alan and Catal proposed approach with naïve bayes(ACN) [23], Alan and Catal proposed approach with random Forest (ACF) [23], our proposed fuzzy majority ranking approach with outliers (FMRT), and our proposed fuzzy majority ranking approach with outliers removal (FMR).

In order to answer research question 1, we look at FMRT rows in Tables 1 to 7. As it can be seen, our proposed model performed almost same as naïve bayes and random forest with slight difference in error, FNR, and FPR values in NASA sets. FNR value in CM1 is improved with 10 and 20 percent compared to naïve bayes and random forest respectively. In Turkish set, AR4, on the other hand, FNR and error rate were increased considerably while FPR value has decreased. According to the results, we can conclude that our proposed method performed as well as two well performed learning algorithm but it does not outperformed them generally, so we accept null hypothesis 1.

Table 1 Results on KC1

Method	FPR	FNR	Error
NB	0.09	0.63	0.17
RF	0.06	0.73	0.16
ACN	0.07	0.16	0.06
ACF	0.01	0.31	0.03
FMRT	0.06	0.74	0.16
FMR	0.03	0.13	0.03

Table 2 Results on KC2

Method	FPR	FNR	Error
NB	0.05	0.53	0.15
RF	0.09	0.53	0.18
ACN	0.06	0.10	0.07
ACF	0.03	0.27	0.06
FMRT	0.06	0.52	0.17
FMR	0.04	0.18	0.06

Table 3 Results on CM1

Method	FPR	FNR	Error
NB	0.08	0.71	0.14
RF	0.02	0.81	0.09
ACN	0.05	0.29	0.07
ACF	0.02	0.42	0.05
FMRT	0.04	0.61	0.16
FMR	0.01	0.37	0.04

Table 4 Results on PC1

Method	FPR	FNR	Error
NB	0.05	0.74	0.10
RF	0.02	0.75	0.07
ACN	0.03	0.46	0.05
ACF	0.01	0.42	0.03
FMRT	0.04	0.80	0.09
FMR	0.00	0.37	0.02

Table 5 Results on JM1

Method	FPR	FNR	Error
NB	0.05	0.79	0.19
RF	0.06	0.77	0.20
ACN	0.03	0.41	0.08
ACF	0.03	0.42	0.08
FMRT	0.07	0.78	0.07
FMR	0.07	0.38	0.12

Table 6 Results on AR4

Method	FPR	FNR	Error
NB	0.06	0.56	0.15
RF	0.06	0.61	0.16
ACN	0.02	0.33	0.07
ACF	0.02	0.37	0.08
FMRT	0.11	0.00	0.09
FMR	0.00	0.20	0.03

Table 7 Results on AR5

Method	FPR	FNR	Error
NB	0.14	0.13	0.14
RF	0.11	0.38	0.17
ACN	0.07	0.00	0.06
ACF	0.14	0.03	0.06
FMRT	0.11	0.50	0.18
FMR	0.14	0.00	0.10

We consider FMR row in Tables 1 to 7 to answer research question 2. Our proposed model performed best in all evaluation criteria in PC1 and AR4. It also performed well in terms of FNR in all datasets except KC2 and CM1, which the FNR rate in ACN were the best. FPR rate in KC1 and JM1 were the best when ACN was used. Error rate was also the best in our proposed method except for JM1 and AR5.As it is also shown in Figs. 3 and 4, our proposed model (FMR) is outperformed other methods in terms of overall error in all datasets except JM1 and AR5. According to the results, we can conclude that our proposed method outperformed other method in terms of error and FNR ratein more than half of the test cases, so we accept hypothesis 2.

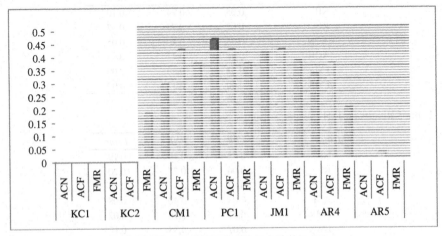

Fig. 3 Method comparisons based on false negative rate (FNR)

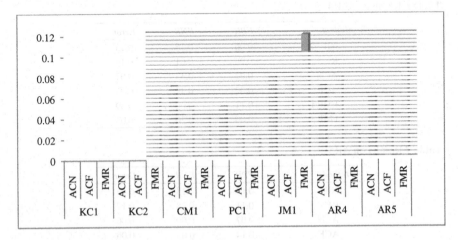

Fig. 4 Method comparisons based on overall error rate

In order to answer research question 3, we look at Tables 8. We should mention here, since Alan and Catal [23] did not evaluate their works based on testing, and training datasets from different software projects, we conducted these experiments according to their algorithms proposed and published in their paper. As it can be seen in Tables 8, our proposed model performed best in terms of error rate in all datasets except (AR4-AR5). It also outperformed other methods based on FPR rate in almost all datasets. FNR rate has not improved based on our proposed method compared to others. According to the results, we can conclude that our proposed method can performed well compared to other methods, so we accept hypothesis 3 as well.

Table 8 Results based on different training and testing Sets

Dataset	Method	FPR	FNR	Error
JM1 for training & CM1 for testing (JM1-CM1)	ACN	0.03	0.41	0.06
	ACF	0.02	0.44	0.05
	FMR	0.02	0.48	0.05
JM1 for training & KC1 for testing (JM1-KC1)	ACN	0.00	0.71	0.08
	ACF	0.01	0.50	0.07
	FMR	0.00	0.63	0.07
JM1 for training & KC2 for testing (JM1-KC2)	ACN	0.00	0.54	0.12
	ACF	0.02	0.44	0.11
	FMR	0.00	0.48	0.10
JM1 for training & PC1 for testing (JM1-PC1)	ACN	0.02	0.39	0.04
	ACF	0.02	0.33	0.04
	FMR	0.02	0.33	0.04
AR4 for training & AR3 for testing (AR4-AR3)	ACN	0.14	0.00	0.12
	ACF	0.09	0.00	0.08
	FMR	0.02	0.00	0.02
AR4 for training & AR5 for testing (AR4-AR5)	ACN	0.11	0.00	0.09
	ACF	0.07	0.00	0.06
	FMR	0.07	0.14	0.09

7   Conclusion

In this paper, we have evaluated the effectiveness of fuzzy clustering based on majority ranking in predicting faulty software module as compared to other famous learning methods. Our proposed method clusters the input data based on available industrial measurement thresholds and then any test data is labeled based on the shortest distance from the modules in three most similar clusters. The overall error rates of software fault prediction approach by our proposed model is found comparable to other existing models and are presented in Tables 1 to 8. According to the results presented in Tables 1 to 7, in most of the cases, FNR and error rates have decreased and the results show that our proposed algorithm works as effective as other software prediction methods.

Acknowledgement. The UniversitiTeknologi Malaysia (UTM) under research grant 03H02 and Ministry of Science, Technology & Innovations Malaysia, under research grant 4S062 are hereby acknowledged for some of the facilities utilized during the course of this research work.

References

1. http://www.pmhut.com/the-chaos-report-2009-on-it-project-failure (retrieved August 3, 2013)
2. Hall, T., Beecham, S., Bowes, D., Gray, D., Counsell, S.: A systematic literature review on fault prediction performance in software engineering. IEEE Trans. Softw. Eng. 38(6) (2011)
3. Catal, C., Sevim, U., Diri, B.: Clustering and metrics thresholds based software fault prediction of unlabeled program modules. In: Sixth International Conference on Information Technology: New Generations, ITNG 2009, pp. 199–204 (2009)
4. Zadeh, L.A.: Fuzzy sets. J. Information and Control. 8, 338–353 (1965)
5. Catal, C.: Software fault prediction: A literature review and current trends. J. Expert Syst. Appl. 38(4), 4626–4636 (2011)
6. Catal, C., Diri, B.: A systematic review of software fault prediction. J. Expert Syst. Appl. 36, 7346–7354 (2009)
7. Evett, M., Khoshgoftar, T., Chien, P.D., Allen, E.: GP-based software quality prediction. In: Proceedings of the Third Annual Conference Genetic Programming, pp. 60–65 (1998)
8. Koprinska, I., Poon, J., Clark, J., Chan, J.: Learning to classify e-mail. Inf. Sci 177, 2167–2187 (2007)
9. Thwin, M.M.T., Quah, T.S.: Application of neural networks for software quality prediction using object-oriented metrics. J. Syst. Softw. 76, 147–156 (2005)
10. Menzies, T., Greenwald, J., Frank, A.: Data mining static code attributes to learn defect predictors. IEEE Trans. Softw. Eng. 33(1), 2–13 (2007)
11. El Emam, K., Benlarbi, S., Goel, N., Rai, S.: Comparing case-based reasoning classifiers for predicting high risk software components. J. Syst. Softw. 55(3), 301–320 (2001)
12. Yuan, X., Khoshgoftaar, T.M., Allen, E.B., Ganesan, K.: An application of fuzzy clustering to software quality prediction. In: Proceedings of the Third IEEE Symposium on Application-Specific Systems and Software Engineering Technology. IEEE Computer Society, Washington, DC (2000)
13. Catal, C., Diri, B.: Investigating the effect of dataset size, metrics sets, and feature selection techniques on software fault prediction problem. Inf. Sci. 179(8), 1040–1058 (2009)

14. Catal, C., Diri, B.: Software fault prediction with object-oriented metrics based artificial immune recognition system. In: Münch, J., Abrahamsson, P. (eds.) PROFES 2007. LNCS, vol. 4589, pp. 300–314. Springer, Heidelberg (2007)

15. Catal, C., Diri, B.: Software defect prediction using artificial immune recognition system. In: Proceedings of the 25th Conference on IASTED International Multi-Conference: Software Engineering, pp. 285–290 (2007)

16. Zhang, H., Zhang, X.: Comments on data mining static code attributes to learn defect predictors. IEEE Trans. Softw. Eng. 33(9), 635–636 (2007)

17. Menzies, T., Dekhtyar, A., Di Stefano, J., Greenwald, J.: Problems with precision: a response to comments on data mining static code attributes to learn defect predictors. IEEE Trans. Softw. Eng. 33(9), 637–640 (2007)

18. Mahaweerawat, A., Sophasathit, P., Lursinsap, C.: Software Fault Prediction Using Fuzzy Clustering and Radial-Basis Function Network. In: Proceedings of the International Conference on Intelligent Technologies, pp. 304–313. InTech/VJFuzzy, Vietnam (2002)

19. Mahaweerawat, A., Sophatsathit, P., Lursinsap, C.: Adaptive self-organizing map clustering for software fault prediction. In: Fourth International Joint Conference on Computer Science and Software Engineering, KhonKaen, Thailand, pp. 35–41 (2007)

20. Zhong, S., Khoshgoftaar, T.M., Seliya, N.: Unsupervised Learning for Expert-Based Software Quality Estimation. In: HASE, pp. 149–155 (2004)

21. Zhong, S., Khoshgoftaar, T.M., Seliya, N.: Analyzing software measurement data with clustering techniques. IEEE Intell. Syst. 19, 20–27 (2004)

22. Yuan, X., Khoshgoftaar, T.M., Allen, E.B., Ganesan, K.: An application of fuzzy clustering to software quality prediction. In: Proceedings of 3rd IEEE Symposium on Application-Specific Systems and Software Engineering Technology, pp. 85–90 (2000)

23. Alan, O., Catal, C.: Thresholds based outlier detection approach for mining class outliers: An empirical case study on software measurement datasets. J. Expert Syst. Appl. 38, 3440–3445 (2011)

24. Rodriguez, D., Ruiz, R., Riquelme, J.C., Harrison, R.: A study of subgroup discovery approaches for defect prediction. Inf. Softw. Technol. 55(10), 1810–1822 (2013)

25. Dunn, J.C.: A fuzzy relative of the ISODATA process and its use in detecting compact well-separated clusters, pp. 32–57 (1973)

26. Bezdec, J.C.: Pattern recognition with fuzzy objective function algorithms. Plenum Press, New York (1981)

27. Promise Software Engineering Repository,
http://promisedata.googlecode.com/svn/trunk/defect
(retrieved August 12, 2012)

28. C. Group. Integrated Software Metrics, Inc. (ISM),
http://innovawv.org/success/ism.asp

29. Abaei, G., Selamat, A.: A survey on software fault detection based on different prediction approaches. Vietnam J. Comput. Sci., 1–17 (2013)

Author Index

Printed in the United States
By Bookmasters